RFID METRICS

**Decision Making Tools
for Today's Supply Chains**

RFID
METRICS
Decision Making Tools
for Today's Supply Chains

William Oliver Hedgepeth

CRC Press
Taylor & Francis Group
Boca Raton   London   New York

CRC Press is an imprint of the
Taylor & Francis Group, an informa business

CRC Press
Taylor & Francis Group
6000 Broken Sound Parkway NW, Suite 300
Boca Raton, FL 33487-2742

International Standard Book Number-10: 0-8493-7979-2 (Hardcover)
International Standard Book Number-13: 978-0-8493-7979-6 (Hardcover)

Visit the Taylor & Francis Web site at
http://www.taylorandfrancis.com

and the CRC Press Web site at
http://www.crcpress.com

Dedication

1. Housman, A. E., To an athlete dying young, in *A Shropshire Lad*, The World Publishing Company, Cleveland, 1932.

Preface

This is a textbook. This book is a systems-centric view of radio frequency identification (RFID). This book is for students of supply chain management, logistics, business management, and project management, and for corporate and military decision makers. It is intended to be used as a textbook in a class on supply chain measurements, with specific focus on the implementation of RFID.

This book focuses on how technology has changed the way decision makers use operations management tools since the computer age and RFID began in the 1940s. Readers do not have to be in a formal academic program of study. However, if you are studying business systems, project management, logistics, supply chain management systems, and globalization of markets, then this book will be very useful. This is not a technical book on the inner workings of RFID. For that, there are several very good selections available for you to purchase.

This book describes how to measure RFID's business use and not make a serious mistake doing so. It is about metrics, metrical units, technological measures and how to use them when making decisions as a decision maker, or how to think about the alternative choices. It is about metrics for RFID. This book is based on 4 years of testing an RFID war game that seems to indicate how decision makers can avoid the pitfalls of the wrong choice.

This book will help you build a metrics frame of reference, help you construct an ongoing metrics laboratory for continuous improvement of your business case, and show you how to run an RFID war game.

This book is for the small business company that has been in business for 1 to 5 years; the medium-sized company that is mature and has a healthy cash flow; the mega-sized company that reaches across the world in flow of goods and services; and the leaders of a state or government investing in RFID technology.

About the Author

Oliver Hedgepeth, associate professor of logistics, is chair of the logistics department, College of Business and Public Policy, University of Alaska Anchorage. His research field spans computer technology, neural networks, RFID, airship technology, and measurement tools. He pioneered artificial intelligence (AI) expert systems for military logistics needs and was the first director of the AI Center for Army Logistics. He began developing computer technology systems in 1967 as a mathematician for the U.S. government, with his first retirement after 30 years as an operations research analyst for the U.S. Army. He holds a Ph.D. and M.E.M. in engineering management from Old Dominion University and a B.S. in chemistry and mathematics from Atlantic Christian College, with graduate research in nuclear engineering from Catholic University of America.

Features and Content

This is not your typical book on RFID. This is a textbook; it is based on what I teach as "dirty logistics." That is, I like to take students and practitioners into the warehouse and show them what a warehouse is actually like, and to try to define a real-world problem for the owner or manager. This book is, then, for those of you who want to know something about RFID, but desire more to make sure you are solving the right problem with RFID. Solving the right problem is more important than implementing RFID the wrong way, or implementing RFID when you really do not need it. More than that, identifying the right problem is the crux of this book.

Chapter 1 is an RFID primer for those who need a refresher on RFID and bar codes. Chapter 2 is where systems thinking is presented. This is not a definitive course of systems concepts, but it is essential that you understand the basics if you are to try to solve the correct problem with RFID. Chapter 3 is where we see some of the many applications of RFID around the world. Some of the unique applications in certain countries are surprising. China and the Russian Far East are showcased, because I teach and lecture in both countries and have for the last 5 years. There are a few surprises in there. Chapter 4 is where I start describing some of the key metrics that will influence your decision to implement RFID. Chapter 5 is where I explore a few new metrics for RFID. There will be more; we all will find them, some more slowly than others. Chapter 6 is where I introduce the RFID war game. This is the heart of the book. The war game has been played for nearly 5 years in Russia, Hawaii, and Alaska. Each time, it has been fine-tuned. It really works to focus your problem if you follow the steps; it is not easy to accomplish. Chapter 7 is a statistical analysis or setup of how to go about analyzing the data that comes from your data requisitions and tracking thousands of records that use RFID tags, compared to non-RFID tracking systems. Finally, there is an Epilogue, where I lay out a few key points from more than 500 articles I have read on RFID over the last 5 years.

Additional material is available from the CRC Web site: www.crcpress.com. A complete set of PowerPoint presentations is available for each chapter. The focus points from each chapter are presented in these 200 slides as a basis of course materials in RFID or as an additional course blending for each instructor's unqiue teaching style. These slides can be downloaded from CRC Press at http://www.crcpress.com/e_products/down-loads/download.asp?cat_no=7979.

Feedback

You can reach the author at afwoh@cbpp.uaa.alaska.edu or at ohedgep477@aol.com or through the publisher's web site.

Acknowledgments

There are many who helped provide information for this book. I would be remiss if I did not let the readers know that many forward-thinking students at the University of Alaska Anchorage, in both the graduate and undergraduate programs, contributed with research in key areas. Kyle Stevens helped with research into how RFID is being used around the world, and Wei Wei Tesch provided valuable insight into China by researching and reading more than 100 Chinese-language web sites on RFID. Don Harman, who opened his air cargo igloo-repair business to the class on RFID. Joe VanTreeck, president and CEO of Matanuska Maid Dairy, who opened his doors and loading docks to our students to experiment with cool chain and packaging options with RFID tags. Bill Tuttle, General William G.T. Tuttle, Jr. (Ret.), who gave me the opportunity to think about this book back in the 1980s and 1990s. Tom Edwards, who has always encouraged me to reach a little further. Gene Woolsey, who does not remember me, but I remember him and his brilliant approach to measuring everything. Robie Strickland, who continues to hound me to complete any writing project. Carol Strickland, who is like a sister, giving me support to continue. Neffie Bennett, whom I call my big sister, and reminds me of family ties, long memories ago. Hallie Bissett, who helps me still try to find new ways of thinking about business success factors. Bear Baker, who prodded me to complete this task. Ted Eschenbach, who actually opened the doors to be in Alaska to conduct this research. Janet Burton, who polished the photos and was always there. Walt Hollis and Wilbur Payne, my two mentors from way back in time. Walt, for always giving me a chance to try something new. Wilbur, for encouraging me to stick my neck out to say something new. Morgan Henrie, who keeps pushing me to write the next article. Many students in my RFID classes. Cameron Perry, who found the hidden treasure of writing; and Candice McDonald, who kept a wary eye and voiced concerns about RFID and Big Brother. Derya Jacobs, who taught me to craft the word.

For Sara, who taught me what it means to be a father; for Will, who showed me how. For Elizabeth, who demonstrated how to write, daily, with fiery editorials when needed. For Ashley, Matthew, Taylor and Olivia, who were always there with encouragement.

Contents

chapter one

RFID *primer for logistics and supply chains*

"There is no value from RFID for us."[1]

Pharmaceutical company representative, May 2, 2006

Introduction

The birth and growth of the computer industry in the United States and Europe, during the decades of the 1950s to 1980s, demanded new tools and metrics for decision making. The computer generated new quantities of data of various types and availability, and produced new information on manufacturing and operational processes for executives, managers, and workers, and project management and analysts for government and military leaders. The computer became a dependent variable in decision making. The rapid growth of artificial intelligence (AI) expert system software, during the 1980s to 1990s, from the United States, Europe, and Asia, started pulling decision-making tools from the custody of the analyst's cubicles into the CEO's and military's senior executive service offices.

This computer machine, from office automation using automatic data processing (ADP) to expert systems, provided a foundation for higher quality manufacturing production and more efficient transportation and allowed just-in-time delivery of goods and services throughout complex supply chains. From my beginning days as a mathematician in 1967, my colleagues and I studied possible uses of this fascinating machine. Decision makers sent us to industry and college courses to study ADP practices and principles. I still have those books. It is refreshing to find passages like "the amazing and wonderful 'world of computers' represents a great step forward in devising better methods to aid man in performing routine and repetitive tasks more efficiently."[2] We could probably replace the "world of computers" with the "world of RFID" as we read the hundreds of trade journal articles on how

RFID will provide benefits and value in performing routine and repetitive tasks more efficiently and effectively.

The ADP text describes how "electronic 'dehumanization' plays a pleasantly constructive role in wide areas of business and, to a limited extent, in other fields. It has taken over the drudgery of routine work by processing data more accurately and quickly than man could possibly do it."[2] The term "pleasantly constructive role" did not go over well with all members of society. If you remember a public television broadcast years later, there was an interview of a man who spoke of turning to a life of crime if computers took his job.

In the 1960s, we mathematicians—the official title for those who later became computer programmers—could see no limit to the use of computer technology. As the ADP book almost gleefully cites, the computer would help humans, at the lowest level of the pay scale, by being a policeman's watchdog, which was the use of computers to store files; the post office and mail transportation, where computers could be used to plan mail delivery routes for trucks; book composition, where a 400-page book could be printed in just under 4 hours, rather than the usual month; automated real estate, where data on houses for sale could be stored and processed using punched paper cards; computerized bread making, with computers being used to check the use of ingredients needed to make bread; banks and computerized checks, where computers are used to process millions of paper checks, stored on magnetic tape with data metrics such as the bank on which the check is written and the individual's account number; automated flight reservations, where computers can confirm a seat reservation, instantly, among over 3000 ticket agents; hotel reservations; air traffic control; and weather forecasting.[3] The creative and innovative uses of the computer in the 1960s continue to evolve as scientists and business decision makers create new uses and new metrics for decision making.

The metaphors we use to distinguish between automatic and electronic have changed as well. Back in the 1960s, the term *automatic data processing* was seen as something different compared to *electronic data processing* (EDP). It was considered that all data processing methods are automatic, meaning more mechanical than electronic. From a dictionary of that era, ADP is defined as, "Data processing performed by a system of electronic or electrical machines so interconnected and interacting as to reduce to a minimum the need for human assistance or intervention."[3] Electronic data processing is defined as, "Data processing performed largely by electronic equipment."[3] From this same era, we used a metric to measure data that was read and transmitted. The metric was the data-transmission utilization measure, which was the ratio of useful data output of a data-transmission system to the total data input."[3] This computer evolution continues with today's focus on RFID data collection and transmission, except electronic and automatic data processing have merged into today's data iceberg within RFID.

The 1990s saw a wave of new information technology and business re-engineering methods, and knowledge management systems that threatened the flattening of organizational lines of communication. Today, with the Internet, GPS, GIS, and wireless e-mail connections, the final roadblocks to more efficient transportation management and planning seem to be at hand—again. This is the realm of the small, passive RFID tag.

RFID tags are the new technical paperclip of data management. The RFID computer chip resembles a small piece of art class glitter. However, that RFID piece of glitter can also be imprinted into a paper label or connected to a metallic ink logo or to a metallic platform, creating an antenna-computer chip assembly. This small computer chip is like the ubiquitous paperclip, securing pieces of valuable information. It stores bits of data, but much more than found on the traditional bar code for product identification.

Unlike the bar code, these chips will talk to your computer, 200 to 1000 times a second, nonstop, 7 days a week, for years in the future, at least theoretically. The voice of these bar code replacements are digital bits that have enough space to store data that are truly interactive; the data can be linked to your company's inventory system to remind you when to move a retail item to the front of the store for a seasonal sale. It can scream, figuratively, to your computer inventory system, from the back of the storeroom, to remind you that the product is about to expire and you had better do something quickly—if you want to sell it. You have a capability with RFID tags to have better inventory and transportation management of those goods.

This wave of RFID technology in the United States is being pushed onto the marketplace by two giant market and political forces: one is Wal-Mart; the other is the Department of Defense (DoD).[4] Both are the largest consumers of American goods, and goods from overseas. Both mandated that the small, passive RFID tag be placed on pallets of goods by 2005 for most, if not all, shipments of goods. This is significant when you consider that Wal-Mart has more than 10,000 suppliers and the DoD has about 42,000; on a good day, Wal-Mart processes more than 250,000 trucks daily, unloading and loading goods from distribution centers and Wal-Mart stores. An economic question many trade articles entice the reader with is, What is the real cost of using such RFID technology, especially on the transport side of supply chain and logistics management? There is no simple answer to this question.

History of RFID

RFID has matured since its birth during World War II. Until 2003, RFID tags were too expensive and too limiting in business applications to be practical for many low-cost commercial items or bulk packaging.

The experimentation with radio frequency and radar before World War II led scientists to research how radio and radar could be used as a means of improving communications. During World War II and immediately following it, the interest in the civilian applications of this technology grew.

But it was not until 1969 when Mario Cardullo had the concept of what we now call RFID. Later, in 1973, Cardullo was issued a patent for RFID. During the 1970s, RFID technology began growing, from electronic license plates for motor vehicles to animal implants in 1979. By the early 1980s, when expert systems were being experimented with for automatic identification of objects, the first toll collection systems for vehicles opened in Norway. During the 1990s to 2000, more than 350 patents were issued for RFID devices. And the formation of the MIT automatic identification (Auto-ID) laboratory ushered in a new era of RFID for logistics and supply chain uses.[5,6,7,8,9]

Today's history lessons with RFID are pilot tests from pharmaceutical companies to milk companies to clothing manufacturers. With more than 50 years of history, RFID is entering a new phase of its existence, brought about by cheaper computer chips, the Internet, cell phone technology, and business process efficiency.

How history will write the story of RFID in the early 2000s will be as interesting as the failed assumptions made on other technologies that impacted business and our daily life. For example, the telephone was thought to have too many shortcomings to be seriously considered as a means of communicating long distances, according to a Western Union memo in 1876. Thomas Watson of IBM said in 1943 that the world market for computers was no more than five. Ken Olson, president of DEC, said in 1977 that there would be no reason for people to have a computer in their homes. The ubiquitous blackboard had a start in the 1850s and was a key innovation that revolutionized education as we knew it then. So, before we get lost in the current history of what we have with RFID for the last 50 years, let us consider what the new history will tell us in the next 5 years, not to mention the next 10 or 20 years.[10]

RFID compared to bar codes

The story of RFID is not all about radio wave technology. It is about how we use this technology versus other technology to identify something, like a pencil, a package delivered to your door, a textbook bought for class, a person entering the Pentagon or an AT&T building. Today, most people in the United States under 30 think it is normal to have bar codes on all packaged goods, like toothpaste or cereal, that you purchase when you go to the grocery or department store. But it has only been about 30 years since Wal-Mart started pushing this technology onto suppliers. Bar code and RFID technology usage is different for different countries. In the Russian Far East, that memory of ubiquitous bar codes would only go back to 2000. Clerks at many Russian stores use some form of handheld or stationary electronic bar code scanner. That scanner reads the bar code on the package, and you see the cost of the item on the checkout display. However, processing an item is still more problematic in the Russian Far East than it is in the West. It could take three clerks to finally check you out of the store for the items purchased. Job security and physical security of cash and

 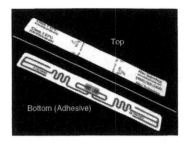

Figure 1.1 Sample of a bar code (left) and an Alien™ RFID tag (right) used for product identification (photo courtesy of Alien Technology, Inc.).

property are still an issue. It is still common in the Russian Far East today to see the actual price written by hand on the item or a printed price affixed to that item.

Most produce, like apples or bananas, does not have bar codes, at least not in the United States. But in China, in 2006, RFID tags started showing up on individual produce items. In the United States, there could be a picture of the item on a keyboard that is selected and then the item is weighed and the price per pound displayed on the computerized checkout machine. Markings of price can be handwritten, computer printed, or wireless, using an RFID tag. The bar code you are most familiar with looks much like the one in Figure 1.1.

The bar code is an automatic identification technology, but so is the digital data encoded in an RFID tag or a reader using radio waves to capture it. When you connect or make a sandwich of a bar code and RFID tag, you create what is termed a *smart label*.

To better understand the RFID tag, let us examine the characteristics of its predecessor, the bar code, in a logistics environment. There are at least eight characteristics that we will use to describe the bar code and RFID tag.

No line of sight

The first characteristic is that the bar code requires the reader to be in the line of sight of the bar code. That is, a person must actually see the bar code, so that person can aim a reading device at the bar code label. How many times have you been at the checkout counter and seen the checkout person looking on the package, turning it over several times in an effort to locate the bar code? Once the clerk sees that bar code, then the clerk has to place it in front of an electronic reader or, while holding a handheld device, scan the bar code. Sometimes, the bar code data is in error, the paper label is damaged, or the bar code is not registering correctly in the computerized cash register. Then, the checkout person has to manually scan, locate, and translate or read the numbers at the bottom of the bar code, similar to the label shown in Figure 1.1. Once the correct symbols or numbers are located, they have to be entered by hand onto a keyboard.

The label-reading process is designed for a person or a machine reader to actually see the image or symbol representing the bar code. When using an RFID tag on a product to identify the product's price, there is no need to see the RFID tag. In fact, there is no need for line-of-sight vision of the RFID tag for either a product or multiple products contained inside a package or container. The radio waves coming from the RFID tag are scanned automatically by a reader, which is placed somewhere on the customer encounter space of the checkout clerk, if there even is a checkout clerk. So, the first and probably most important characteristic of RFID tags is that they require no line of sight to be used, whereas bar codes demand line of sight.

Longer read range

The second characteristic is related to the first and is the read range of both the bar code and the RFID tag. The bar code has a limited read range measured in inches or fractions of an inch. The RFID tag read range metric is determined in feet and, depending upon the situation, can be read from 5 to approximately 30 feet (10 meters) between the product with the RFID tag and the reader. This read range is essential for a business case considering the use of RFID tags compared to the line-of-sight bar codes. Currently, when a trailer enters the loading dock area at a warehouse and the forklift driver begins unloading pallets of goods from that trailer, each pallet needs to be read visually to match the bill of lading or airbill or manifest. Some cargo items are obvious by their size and shape; a generator fastened to a wooden pallet looks like a generator. A huge tire looks like a huge tire. But the bar code or shipping information on the side of the generator or tire still needs to be entered into the inventory system.

Figure 1.2 shows a cross-docking facility or warehouse where goods have been taken from several trailers on one side of the warehouse, stacked into areas or zones being prepared to be shipped out to another trailer on the other side of the warehouse. Notice the white labels affixed to each container or specific item. It should be obvious that the large tire is a tire, but where does it go, where is it coming from, and when does it need to be somewhere besides waiting in this warehouse? To find out what is on the floor of this warehouse, someone has to walk around each container or item and record the data on those white pieces of paper you see on each container. This could be done by a dock worker using a clipboard and pencil, or a handheld computer, entering the data manually, or scanning with a line-of-sight reader. The bar code reader requiring the line of sight for recording its data has to be brought within inches of the shipping label. And for the boxes that have individual labels and are part of a shrink-wrapped pallet, each box has to be broken out of the pallet and read, or at least those boxes that are in the middle of the pallet of boxes do.

If each item in this warehouse had been also labeled with an RFID tag, there are several possible scenarios for automatically reading each item you see on the floor. One scenario is to have an RFID reader attached to the top of the bay door, where you see the back of the trailer is located. As the forklift driver goes into the back of the trailer to pick up a load, and then backs into

Figure 1.2 Goods stored in a cross-docking facility.

the warehouse with that load, every item that he or she has picked up on that pallet is instantly read and recorded in an inventory system by date and time in hours, minutes, and seconds. Each box that is blocked from human sight in the large shrink-wrapped pallet is read the second the forklift driver enters the warehouse dock door threshold. The distance between the reader and the boxes that are read and (hopefully accurately) recorded would be in the range of 5 to 8 feet, from experience.

Static data entry

The third characteristic is whether or not the bits of data programmed on the tag can be changed. Going back to Figure 1.1, the bar code data on that piece of paper or plastic or etching on a metal part have been transcribed once. The numbers "14141 00011" are static; they cannot be modified once printed on a label. To change the number would require a new bar code label; a bar code is read-only. However, for the RFID tag in Figure 1.1, data can be written several times or modified during its transit along a supply chain; an RFID tag is read and write and read again, up to 100,000 iterations of changes, theoretically.

Data volume

The fourth characteristic is the amount or number of bits of data that can be stored on the RFID tag and the amount of data stored on a bar code. A sample bar code and RFID tag are shown in Figure 1.3. The bar code number

0 1280-53089-2 2	41.405C7A3.10F3T3.430SPE12A

Figure 1.3 Amount of data that can be stored on a bar code (left) and an RFID tag (right).

is one from an item purchased at a music store. That bar code is made of different sizes of lines and spaces, each configured in parallel. This configuration can be interpreted as numerical or alphanumerical symbols. There are usually 13 digits for a bar code, but there are at least 10 different types of bar codes. The digits represent the country, the company manufacturing the product, the manufacturer's product item number, and a check bit or digit. The RFID tag number shown in Figure 1.3 is similar. The first two digits represent the product version. The second sequence of alphanumeric characters represents the manufacturer. The third sequence of alphanumeric characters represents the goods or product that has been manufactured. The fourth sequence of bits is the serial number for the product.

By inspection of the numerical difference between the symbol characteristics of the bar code and the alphanumeric characteristics of the RFID tag, you can interpret that the number of products and items that can be recorded is significantly (at least more than 10%) more than those used by the bar code. Some RFID tags can contain up to 1 megabyte of memory (1 million characters), although most tags only contain a small fraction of this memory, perhaps as little as 64 bits, which is 5 times more than recorded on a bar code.

Identify more items

The fifth characteristic is that the RFID tag identifies the product and item and possibly much more, depending on the user's needs, whereas the bar code identifies only the product. In the complex world of today's competitive manufacturing, logistics, and supply chain global markets, the ability to track and trace not only the product but the item number, and possibly a history of birth to death movements, adds a new metrical dimension of in-transit visibility to the item.

Retail products and items are not stationary. While they may sit on a retailer's shelf waiting for a customer to pick up an air filter, that air filter has traveled a long way to get to that retail shelf space. The logistics and supply chain threads for a product can be described from its raw material to final product that is packaged and shipped to that shelf space. With emphasis today on environmental impacts of manufactured goods, the ability to identify a used part for recycling is gaining legal momentum. In some countries, certain products must be able to be tracked from raw material to end product to final disposal back into the raw material from which it was made. RFID and bar codes can both help in this regard. But which is best used for such cradle to grave to cradle process? Only time and experimentation will tell; this could be yet another difference between the two types of tags.

Simultaneous data capture

The sixth characteristic is that most RFID systems can simultaneously identify and capture data from multiple tags within range of the antenna. With bar codes, only one bar code tag can be read at a time. However, as we enter the area of thousands of inventory items that need accurate identification, this process can take many hours or days with traditional bar code systems. Experiments with RFID tags seem to indicate the possibility of simultaneous reading of hundreds of boxes at a time, within 1 or 2 seconds.

Rapid read rate

The seventh characteristic is correlated with the sixth characteristic, in that RFID tags can be read very rapidly. RFID readers are capable of capturing tag identification codes at a rate of up to 1000 tags per second, theoretically. Although the theory and principles of physics might indicate many reads per second, experience has shown that there are other factors that could limit this rapid RFID read rate. The collision of RF signals from different areas, multiple sources, how the software works at filtering the signals, the placement of the antennas, and the training of the staff all could contribute to less efficient read rates. Be careful of comparing laboratory read rates with those made in a cross-docking facility, constructed of tons of concrete and steel, which can cause distortion in RF.

Durable for harsh environments

An eighth characteristic of RFID tags is that they can be encased in hardened plastic coatings, making them extremely durable and able to be tracked through harsh production or transportation handling processes. They can be read through grease, dirt, and paint. The RF signals are capable of traveling through a wide array of nonmetallic materials. Bar codes, on the other hand, can be torn by harsh contact or abrasion, whereas a hardened platform can extend the life of an RFID tag. Bar codes can be encased in transparent plastic or glass, as in the case of many warehouse operations.

There is a word of caution concerning these eight characteristics showing the differences between RFID and bar codes. It has not been determined that bar codes will be replaced by RFID tags—not by any business case or research or massive application of RFID tags. We have compared RFID tags to bar codes because bar codes and the inventory system that has evolved over the last 30 years will have to be modified when RFID systems are introduced. My pack of number 2 wooden pencils, which I use when writing notes on a paper pad while reading about RFID use from the Internet, indicate that it is possible that the business case of bar code versus RFID may become bar code plus RFID for a long time to come.

How does an RFID system work?

The RFID system uses a tag, which is a paper and metallic label, composed of a microchip with a coiled dipolar antenna. The label combination may be covered with a plastic protective cover, which can be transparent or not. This RFID label or tag is affixed onto a container or package or onto an item, such as a book or CD. This tag communicates with an interrogator or reader, which also has an antenna. The reader transmits electromagnetic waves that form a magnetic field, which activates the antenna on the RFID tag. A passive RFID tag draws power from this magnetic field and uses it to power the microchip's circuits. The chip then modulates the waves that the tag then sends back to the reader, and the reader converts the new waves into digital data. The reader interprets the tag data and communicates this data to a computer. Also, computer middleware is used to process this data. This data are then stored in a computer memory for processing or interfacing with other software inventory systems or analysis systems. RFID uses the low end of the electromagnetic spectrum. The waves coming from readers are no more dangerous than the waves coming to your car radio. This process is shown in Figure 1.4.

Today, these RFID tags and bar codes are being used together. Before the bar code is brushed aside, the smart business decision seems to be using both the old technology and the new technology together. When you get your luggage tag at the airport, check it out carefully. It may be a sandwich of two pieces of paper, with the inside filling a passive RFID tag and the outside a standard bar code, as shown in Figure 1.5. Actually, the luggage tag will probably have five copies of the bar code on the outside of the luggage tag; if an RFID is also used, there will be one passive RFID tag in the middle of that label. So, already, we can see maybe another characteristic difference between the bar code and RFID tag. How many bar code tags are used in combination with passive RFID tags?

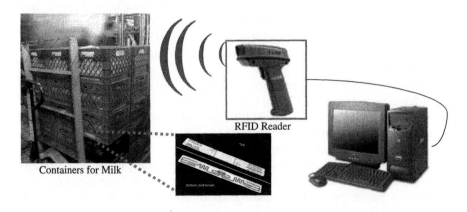

RFID Reader

Containers for Milk

Inventory Control

Figure 1.4 Process of reading RFID tags placed on a container.

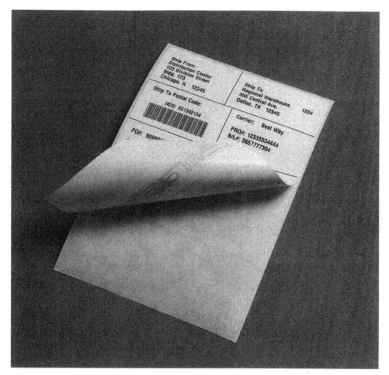

Figure 1.5 Standard shipping label with bar code plus the passive RFID tag inside (photo courtesy of Alien Technology, Inc.).

Information from the hundreds of trade articles, and experience with the bar code and passive RFID tags, does suggest that the use of bar codes and passive RFID tags will continue for some time. There does not seem to be a business case, yet, where one can afford to replace the bar code throughout the entire supply chain system.

If you want to go into the RFID business or experiment with this technology, it is not that costly. The price for a complete kit with reader, antenna, and sample tags runs from several hundred dollars to about $5000. A sample kit is shown in Figure 1.6.

However, a word of caution: This RFID technology is more about data, the business process, and systems thinking than it is about the technology. This holds true for bar code or RFID. Just purchasing the equipment is one small part of a multiple set of solutions with RFID. The other part of the solution set is to analyze and examine your business process of manufacturing or movement of goods, and the potential impact of updating your computer inventory spreadsheet or database if you incorporate RFID. Additional software to interface with your current inventory system is required if you use RFID.

Besides the test kit shown in Figure 1.6, you can attend special 2- or 3-day boot camps at various technology centers, such as Alien Technology,

Figure 1.6 Sample RFID kit that you can purchase for your own small pilot tests (photo courtesy of Alien Technology, Inc.).

Inc. in San Jose, CA, or others at various universities around the country. The cost for a typical boot camp can cost up to $5000 for a few days. But with more university courses becoming available, the cost to learn about and experiment with this technology is becoming more affordable.

The use of these tags is still in the experimental stage, for the most part. Trade magazines and newspapers are full of stories of such pilot tests. Even the ubiquitous Wal-Mart and the Department of Defense are conducting pilot tests, but with a price tag that spans millions of dollars. One example of how a test is conducted in a warehouse is shown in Figure 1.7.

The figure shows a doorframe structure, around which are placed four white RFID readers. Four were chosen to ensure the signal from the RFID tags would be read when the pallet passes through the frame. In the background is a worker about to enter the frame with a load of boxes, each of which has a passive RFID tag affixed to the outside of the boxes. When the pallets are pulled through the frame, each tag registers that a case of beer has just entered the dock doorway and is going onto a waiting trailer for delivery to a customer. Similar antenna designs are being experimented with in pilot tests at warehouses across the country. One of the implementation issues, which affect the RFID metrics, is how many antennas are necessary to ensure a 100% accurate read rate, another RFID metric.

The figure also shows this frame as a temporary structure. In this pilot test, the worker, who happens to be Steve Brown, cofounder of the first RFID

Figure 1.7 Pilot test of placing RFID tag readers around a frame where a pallet of boxes with RFID labels affixed to them can enter.

company in Alaska, Nano-Logistics, LLC, is pulling the pallet. He will pull this pallet slowly through the frame so as not to hit the frame or one of the four white squares, which are the antennas. Under normal warehouse operations, the worker would be moving as fast as he or she can to get the cases to the other side of the warehouse or to load onto a waiting truck. Time equates to money in the transportation business. Under normal circumstances, the probability that the worker would hit the sides of this frame is significant; even a probability of 1% could be crucial when a 100% read rate is needed for a 24-hour operation, moving thousands of boxes per hour. The first time the driver hits this frame, down goes the frame, or off comes the antenna or antenna cable connections. If you have ever been in a cross-docking warehouse, you know that the forklift drivers take great pride in going as fast as they can, without hitting something. However, the marks on the walls tell another story. So, you will have to reconfigure your workspace as well as the process of counting each case that is leaving the warehouse.

Active versus passive tags

RFID tags are either active or passive. The active tags are considered transmitter tags. They have been around since the 1940s. Active tags have a battery for a source of power to store data or to transmit a radio signal with the data. As such, they perform similar to your portable radio and can be read over a great distance, with a limit of over 100 feet (30 meters). Active tags are significantly more costly than passive tags. The active tag is usually encased in a hard shell for protection, because these tags are used in rugged environments. A sample of an active RFID tag is shown in Figure 1.8. These active tags are used for a variety of purposes, such as tracking the time and temperature of fresh Alaskan salmon. The tags are placed into a box of salmon that is to be kept at a constant temperature near freezing for the journey from the ship to a variety of transportation systems: barge, truck, or airfreight.

Other active tags might be attached to boxes of cargo destined for a war zone for the military. They need to be able to withstand pressure, extremes of temperature, and shock. The cost of active tags varies but ranges from about $30 to more than $100. Compare these costs to passive RFID tags, which cost from 20¢ to 30¢. Compare to the paper bar code, which costs less than 1¢, and you have yet another difference between bar codes and RFID tags.

The passive tag, shown in Figure 1.1, can also be encased in a hard plastic shell, as shown in Figure 1.8. The passive tag has no battery. As shown in Figure 1.1, the chip is about the size of a grain of pepper and is usually at the center of a bipolar antenna array. This chip receives energy from the reader, which also transmits a signal, described earlier. The power of the passive tag comes from the conversion of the RF power into a DC current. The passive tag needs to be closer to the reader than the active tag. Experiments and operational usage indicate that placing a reader within about 5 to 8 feet is sufficient to produce the required data transfer. However, they

Figure 1.8 An active RFID tag.

 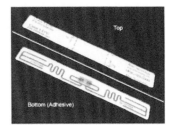

Figure 1.1 Sample of a bar code (left) and an Alien™ RFID tag (right) used for product identification (photo courtesy of Alien Technology, Inc.).

Figure 1.2 Goods stored in a cross-docking facility.

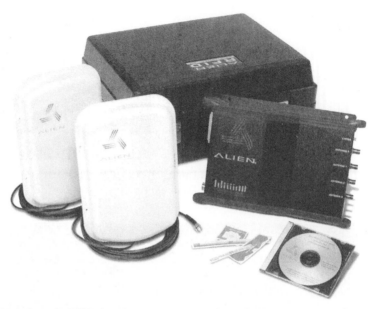

Figure 1.6 Sample RFID kit that you can purchase for your own small pilot tests (photo courtesy of Alien Technology, Inc.).

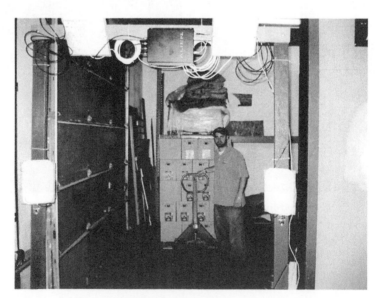

Figure 1.7 Pilot test of placing RFID tag readers around a frame where a pallet of boxes with RFID labels affixed to them can enter.

Figure 1.9 Container ship showing typical containers that could use active RFID tags.

Figure 1.9 Container ship showing typical containers that could use active RFID tags.

can be closer. On occasion, readers have been successful out to 30 feet, but these are in laboratory conditions.

The cost of the passive tag is significantly lower than an active tag. In 2004, the cost of an active tag as shown in Figure 1.1 was around 50¢. Due to mass production changes and expected volume of use, this cost is expected to be around 5¢ by 2008 or 2009. Such economies of scale for cost make the passive tag very attractive for containers of products, but not necessarily on each individual item in the container. But this usage is still not conclusive and needs more research and operational use.

One of the key areas in logistics where active tags are being used is in shipping of cargo entering marine ports. With the millions of containers entering U.S. marine ports each day, the current inspectors cannot open each container to inspect and compare the contents with the manifest. Figure 1.9 shows a TOTEM container ship at the Anchorage marine port facility where the containers are driven off the ship and parked in a waiting area. RFID tags could be used to identify what is in each container and where the container has been parked after leaving the ship.

Why the interest in RFID?

The U.S. Department of Defense and Wal-Mart have mandated the use of passive RFID tags from their suppliers. Although there are mandates for specific suppliers to the DoD and Wal-Mart to be compliant today (they actually started in January 2005), the reality of operational cost and impacts to supply chains and logistics inventory management systems is still not known. But this action is similar to the action made by Wal-Mart nearly 30 years ago when it started pushing bar codes on its suppliers. It was a slow process for manufacturers to adopt bar codes. It will be a slow process for RFID tags as well. During 2005 to 2008, there will be a series of operational tests conducted by Wal-Mart and the DoD to fine-tune the operational needs, test the procedures for use, and determine how best to measure a return on investment.

Alan Estevez, Assistant Deputy Undersecretary of Defense for Supply Chain Integration, stated that for DoD, "The focus on our RFID implementation approach and policy is with the warfighter in mind."[11] He labels RFID as a disruptive technology, in that the use of RFID disrupts the old way of operations in logistics and supply chains and business in general. Estevez said he believes the use of RFID technology in the military supply chain can optimize that supply chain, but only if RFID is adopted at all levels of the military business process. He said, "The way we fight wars is changing. The DoD needs to keep abreast of changes in the way we do logistics in order to maintain the support that our forces deserve, and we believe that the use of RFID technology is critical to doing so."[11] The military approach to using RFID is not new. The military has experimented for years with new computer technology and business re-engineering methods to make the supply chain for the military more efficient. In the 1980s, the use of expert systems as part of the DoD initiative with artificial intelligence software demonstrated that smart software could help pull needed logistics information from different sources into more useful forms. But those were days of experimentation. They laid the foundation for a new way of thinking that had to wait for the current low cost, mass assembly, passive RFID technology breakthroughs to occur. Figure 1.10 shows what concerns the military when logistics visibility goes wrong.

The initial Wal-Mart business venture into passive RFID in the next few years will probably see such tags on the case or carton or pallet of goods and products being shipped. This is mainly due to the cost of the tags. While passive RFID tags are more than 5¢ each, we can expect that such tags on

Figure 1.10 Logistics visibility in the desert.

individual items will be rare, unless the item cost is high enough to absorb the tag cost, without resistance from the consumer. For example, although there is a bar code on a box of toothpaste, there is not an RFID tag. The cost is too high to pass on to the consumer. But for a case or carton or pallet of toothpaste, using such a tag while the product enters the warehouse or stockroom seems reasonable—for now.

Passive RFID tags are also easy to read on conveyor belts. However, the change in an ongoing process is an extra expense, and exactly how much it costs to change the conveyor process to add such tags may slow down some business decisions or processes. A tag could be affixed to the plastic shrink-wrap of some cartons; however, an unresolved issue remains: where to put the tag on the shrink-wrap and how much it will cost to change the process. Adding such tags by hand is not cost effective.

These problems are what the DoD and Wal-Mart hope to solve in the next few years. But others are working hard to understand the role of passive and active RFID tags and systems. They include Gillette, Johnson and Johnson, Kraft Foods, Pepsi, Coca-Cola, Kodak, Kimberly Clark, Pfizer, Home Depot, Target, and UPS, and the list is growing daily.

The pharmaceutical companies are working toward the use of RFID tags to protect the security of drugs that you buy on a daily basis. However, it is interesting that the chapter started with a quote from an executive from a pharmaceutical company attending an RFID event on May 2, 2006. The comment was made to a group of people who are experts in RFID and represented universities working to promote proper use of the technology.

Currently, your bottle of pills is controlled by the use of a bar code that is used to classify one type of pill. A company may have a million bottles containing that pill, but the same bar code is used on each of the millions of bottles. As RFID enters this market, the plan is or seems to be to use a unique RFID for each bottle. The result will be a combination of bar codes that describes that this bottle contains some pill, but also an RFID tag that says this is where the bottle came from, here is where it was last located, and here is where it is supposed to be, with the user's name or identity. A lot more information is given, but for safety and security reasons, it is a necessary step. Compared to the bar code, this could possibly be another difference with the passive RFID tag, but this difference is a business decision to give unique identifiers to each and every item.

Most of this new effort in the pharmaceutical industry is from the post-9/11 security concerns about cross-border counterfeit pills and the possibility of a terrorist attack through contamination. The pharmaceutical industry is extremely vigilant in security and safety; however, RFID provides an opportunity to be not only more vigilant, but also more competitive.

There will be growing interest in the possible uses of RFID in the coming months and years, especially following the mandates of the Department of Defense and Wal-Mart.

Current applications for RFID

Applications for RFID tags have been around for years. What is changing, however, are the applications with passive RFID tags to comply with the mandate for Wal-Mart and the Department of Defense for their major suppliers. This is where most of the newspaper and trade magazine articles are concentrating their fears and desires and trying to answer the question of applications to meet this mandate. But while the suppliers are looking for the best-case scenario to meet the current short-term demands, there are plenty of examples of applications of the passive RFID tags that can help provide evidence on applications. There are, for instance, smart cards used for public transportation in different countries, but they are not so widespread in the United States. There are plenty of examples of transportation smart cards or tickets being used for public transportation with the rail systems in Europe. However, not all countries in Europe subscribe to the use of such smart tickets, or if they do, to the same characteristics. Thus, one ticket does not make possible a train ride all across Europe, not just yet. But this will probably happen over the next few years.

Access control into public and government buildings is already underway in the United States and has been for some time in Europe. In the United States, especially since 9/11, the use of passive RFID tags in personal identification badges is commonplace. Around Washington, D.C., you see this everywhere. You cannot go into a federal building without at least one or more identification badges, one of which is RFID. In some cases, there are people wearing a separate RFID ring on their shirt for entry into yet other layers of secure buildings.

RFID tags have been used for animal identification for years, but they are usually active tags—with a battery. These tags, along with their cousin the bar code, help identify the animals from birth to death. When the animals feed, the computer knows exactly what to feed the specific cow. When a cow gives milk, there is a reader nearby to monitor the milk quantity along with the specific animal.

Another security RFID is in the tracking of humans for the justice departments of the world. Many people do not spend time in jail, but instead are confined to their homes, wearing active RFID tags, often with GPS sensors to track their locations in real-time.

Perhaps the most noticed and written about use of RFID tags since 9/11 is in the container field. With millions of containers entering the marine ports of the United States each year, only a few percent are actually checked for their contents. The use of RFID tags from the point of origin, say from Asia or Europe, into the port of entry in the United States or Canada is receiving a great deal of attention from the Department of Defense and Department of Homeland Security. This is one area that still remains to be tested, and the proper process for use must be identified.

Sporting events have been using passive RFID tags for some time. In Germany, passive RFID tags replaced the paper ski lift ticket for skiers. In the

United States, runners or racers now have the option, in many cities, to wear a small medallion on their running shoes. This medallion is really a passive RFID tag that triggers readers that are placed under mats, over which the runners start and finish the race. Thus, no longer are those runners penalized for being in the back the pack of 10,000 runners. The last runner in the starting lineup can now take the several minutes it takes to reach the start point, knowing that the minute they cross the line, their individual time starts.

The automobile industry has been a leader in using passive and active RFID tags for some time now in the United States and in Canada. The assembly line is now more just in time than ever before, with the parts like seats matched to the body of the car or truck, weeks in advance, and matched perfectly when the seat arrives just in time for placement in the one car or truck for which it is meant.

One of the areas for use of RFID tags is in the medical or health care field. There are some pilot tests under way to tag patients in an emergency room, to keep track of them. But there are also privacy and ethics issues that have yet to be solved. Until these ethics questions are answered, the use of RFID for patients may be slow in coming or, at least, will be stalled for some applications.

RFID is but one auto-ID system

The interest in RFID is paralleling a process of business and government trying to use different kinds of automatic identification or Auto-ID systems. The bar code and the RFID have already been mentioned. They are but two that belong to the Auto-ID family. Another Auto-ID system is optical character recognition, or OCR. This system has been around since the early days of computers as well and started to be used in the 1960s, especially with the military. OCR was an attempt to read characters or letters and numbers the way a human would read them. In the 1970s, I worked with the Army to use OCR for a variety of projects. One was the reading of news about soldiers getting promoted or other good news that could be sent to their hometown newspaper. OCR was seen as a means to read vital information into the computer and transmit that information to those who may need to read it for decision making. However, the early attempts saw massive failure rates of 70% for the first few years. Today, OCR readers are in use in banks (look at your check's bank routing number and the special font used) and other high-value production facilities. However, OCR never became as popular or as successful as bar codes due to the high cost and the need for special readers.

With the advent of post-9/11 laws and regulations governing security at the nation's borders and Customs, another form of Auto-ID, biometrics, has become prominent. If you fly into or out of the country or if you work for the U.S. government, you are all too familiar with matching your fingerprint with an electronic scanner, or looking into a retina or iris or eye identification camera, or speaking your name into a voice reader. Such security systems have been around since the 1980s, when the first commercial uses of what was then called artificial intelligence or AI systems were experimenting with how to uniquely distinguish one human from another. Today, when you enter a secure facility, you may need to use a biometric Auto-ID encounter as well as a photo badge that also has a passive RFID tag on the badge.

Another Auto-ID system that people seem to take for granted is the smart card. This card is your credit card. It stores data in a magnetic strip about who you are and your account number. Many smart cards are also being manufactured with a passive RIFD tag sandwiched between the layers of the plastic. You will know you have a smart card if the card requires you to pass it in front of a reader before it lets you into a building or into the subway station, or to purchase a pair of pants or shoes—something that may become common in a few years.

RFID is not new

The history of technology is full of stories about how people rejected new technologies for one fear or another. Blackboards were introduced into the U.S. school systems about 150 years ago as a new way to teach. Typewriters caused panic among those who wrote orders for the generals in the Army. Television was a great experiment, but was thought to have no practical value, as was the telephone. The Luddites in London smashed textile machines that threatened jobs by automation. But today, RFID seems like it was just found.

References

1. Ronchetti, Mike, of RFID Complete, personal interview while attending an RFID conference, May 2, 2006 Las Vegas, Nevada.
2. Awad, Elias M., *Automatic Data Processing: Principles and Procedures*, 10th ed., Prentice-Hall, Inc., Englewood Cliffs, NJ, 1966, Chapter 1.
3. Sippl, Charles J., *Computer Dictionary and Handbook*, 1st ed., Howard W. Sams & Co., Inc., Indianapolis, 1966, Preface.
4. Heinrich, Claus, *RFID and Beyond*, Wiley Publishing, Inc., Indianapolis, 2005.
5. Lahiri, Sandip, *RFID Sourcebook*, IBM Press, Upper Saddle River, NJ, 2005.
6. Kleist, Robert A., Chapman, Theodore A., Sakai, David A., and Jarvis, Brad S., *RFID Labeling: Smart Labeling Concepts & Applications for the Consumer Packaged Goods Supply Chain*, Printronix, Irvine, CA, 2004.
7. Sweeney II, Patrick J., *RFID for Dummies*, Wiley Publishing, Inc., Hoboken, NJ, 2005.
8. Shepard, Steven, *RFID: Radio Frequency Identification*, McGraw-Hill, New York, 2005.
9. Palfreman, Jon and Doron Swade, *The Dream Machine: Exploring the Computer*, BBC Books, London, 1991.
10. Hazelwood, Lynn, RFID in the Department of Defense: The bottom line, *RFI Defense*, Supplement to Vol. 62, No. 1, *Defense Transportation Journal*, 24, 2006.
11. Estevez, Alan F., Using RFID to Optimize the DoD Supply Chain, briefing topic from IPT Conference, August 6, 2003.

chapter two

Systems concepts
of using RFID

"The flow of ideas from one field into another often takes curious and ambivalent paths."[1]

Loren Eiseley

Marks & Spencer, a well-known British retail chain, used what we call the systems approach back in the 1930s. It designed and tested its products before management decided to sell them. It also used only one manufacturer to produce each product. It worked closely with that manufacturer so that the products, whether textile or food, were produced at the right time and right price to sell to the customers. It was also a pioneer in forecasting methods. In the 1990s, when Peter Drucker wrote about Marks & Spencer, he indicated that the systems approach was still rare in the contemporary manufacturing environment.[2]

This approach to business was not necessarily technology based, but systems based. Contrast Marks & Spencer with Henry Ford, who built the Model T automobile. He wanted to control the steel mills, the glass manufacturing, the rubber plantations, and rubber supplies, and considered building gas stations and service centers to repair his cars after people bought them. What Henry Ford built was not based on a systems concept, but one of ownership; his financial power held this conglomerate together and made it function as a well-oiled machine. But, eventually, this machine became too large, with too many moving parts that worked against the personal desires and drives of a single person.

Henry Ford thought he could control a system, his system of thousands of components, each working toward his common goals, rather than the components' common goal. This was not a systems approach. In fact, today's manufacturing facilities are not controlled at one point. The component parts of today's manufacturing systems are independent parts, contributing to the overall goal, whatever that might be. The system of today still has one very common element with supply chain studies and business practice: the customer. The planning for any system that is successful today starts with the delivery to the customer, and the planning to get to that point starts a backward planning exercise. Drucker in the 1990s said, "The new system sees the plant as little more than a wide place in the manufacturing stream."[2]

We now will add the language of systems, before we jump from theory to applying RFID technology to boxes, pallets, or igloos, and before you start using your traditional metrics to show how successful you are.

The study of systems had its birth in World War II, when the operations researchers (OR) began to experiment with finding order within chaotic environments. When systems are observed or recorded over time, the set of continuous states is called a dynamic system, and the system exhibits dynamic behavior. Systems can be categorized and analyzed as probabilistic, stochastic, random, simple, complex, or self-regulating. The dimensions of systems, the pace of their values, and their control is the foundation of understanding how we apply system metrics.

Decision makers face many problems of control and communications with RFID technology, not much different in the complexity of the system that Henry Ford tried to control. By understanding the RFID system variables, we will begin to peel back the suspected problems of RFID and uncover the applicable problems to be solved with RFID, which is the purpose of this book.

It is hoped that you started this book with a problem to solve using RFID. It is also hoped that when you finish this book, you either know you have the right problem, or you have thrown that initial problem far away, or you know you almost see the problem.

Systems

A system is "a group or set of related or associated material or immaterial things forming a unity or complex whole."[3] A large supply chain for a global retailer such as Wal-Mart constitutes a group or set of material things, such as retail stores and distribution centers. *Large* is defined as more than 5000 retail stores. A smaller group or set of material things could be Matanuska Maid Dairy, from Alaska, which produces milk only inside the borders of Alaska. Or a smaller viewpoint of a group could be the single Wal-Mart store in Ashland, VA, which is part of that larger system we could call the Wal-Mart system. All three examples share common attributes or have material things that could be considered a group or set of those things. But what are the events or processes that cause these systems to be related? To say that all three systems form a unity or complex whole is also problematic. This definition is from a dictionary. As such, it provides a general description for those who need to have a generalized understanding of the concept of system.

A system can also be described as "an object in which variables of different kinds interact and produce observable signals."[4] This definition has parameters similar to the first definition. However, it is more mathematical or statistical in its meaning. In this definition, the Wal-Mart store, the over 5000 Wal-Mart stores, or the one Matanuska Maid Dairy are considered objects. The objects are then described as mathematical variables.

A significant part of this definition is the term *interact*. If you consider all the variables in a system as a road going in a straight line, nothing unusual happens to that variable. However, when you have these roads cross each other, creating a crossroads of the flow of information, a crossroads of credits and debits, and a crossroads of customer to retailer interaction, then the information has interacted for different paths, and new information is formed. Only at the crossroads of the supply chains can new information be created, new products designed, and new problems emerge.

A system can also be thought of as "the coherence of a number of entities called parts of that system."[5] But what is the coherence factor of these entities? The flight of an Air China cargo jet from Shanghai to Anchorage to Boston represents such as system. It is a system of systems. There is the macro system of cargo movement from China to the United States, and the return flight back to China. There are the jet engines that power the huge planes. These engines have a coherence of parts that move toward a stated purpose. Likewise, the jet moving along a specified path from China to the United States is a system of parts, crew, fuel, cargo, and communications devices. However, all these parts, or entities, or things have no meaning until we give them purpose. This purpose is granted by the decision maker at various levels of an organization that interacts and tries to manage its view of the system.

Stafford Beer wrote that one can recognize a system by observing that system within three stages of its life. First, we make the assumption that a unique relationship exists, where a collection of things that can be considered assembled, or an assemblage. Second, we observe a pattern of activity emerging from this assemblage. Third, we observe or assume a purpose to this pattern.[5]

All three stages have a metrical standard, which could be counting the collection, ordering the collection, weighing or sizing the pattern over time, and checking the purpose of this assemblage with that of the organization. The jet, the jet engines, the cargo, the crew, and the flow of goods along a geographical path all are systems. You do not need to see them. The decision maker, however, has to see them, as well as believe in them; otherwise, the system's purpose to that of the organization's mission is false.

The decision maker, from the loading dock supervisor to the CEO, is concerned with the model of their system's view of the organization. Managing the various levels of viewpoints, over a timeframe, with specific and vulnerable operational assumptions, is the task laid out in this book.

The decision makers can separate their subsystems of the large system into performance variables and metrics. The statistical world of industrial engineering, operations research, systems analysis, and the latest incarnation of all of that, Six Sigma and the Balanced Scorecard, shows management's intent to understand assemblage, pattern, and purpose.

Taking the parts of a system and measuring them by some performance indicator is done all the time, in large and small organizations and in democratic or nondemocratic governments. We warn people not to suboptimize their system, that is, not to make the flow of goods on assembly line 13 produce error-free metal plates, without understanding where those plates are going next, and where they came from in the first place. To pay attention to one part of the assembly line at the expense of the other parts of that system will help cause the scrap pile to increase in quantity of failed parts. So, we can look at the parts of the system, like my cardiologist looks at my body's physiological and chemical parts, or like we look at the China Air cargo supply chain.[5]

To view a system, we can start at 30,000 feet in the air as you fly to your next business meeting, where you could look down at an interstate highway across Kansas. There you could see cars and trucks, looking like little gray or black dots or dashes riding along a black ribbon of highway. The details of those cars or trucks or tractor-trailers are lost from your view. But you can see part of the highway that connects the cities with freight movers and people movers. Now, compare what you would be seeing from the car you are driving or the tractor-trailer in which you are riding. In both instances, you see time differently. Now, place yourself in the cross-docking warehouse where that truck is being unloaded of 500 boxes that it was carrying along that stretch of interstate highway. The 30,000-foot view is down to 30 feet, but still part of that 30,000-foot view. What is your role in the warehouse compared to that in the plane?

A system is a bunch of parts where each part interacts with each other part. This is a very simplistic view of how humans and machines work together toward some common goal.

Think of the daily process of getting up in the morning and the dozens or hundreds of trips you take from one point to another, just in the house; this comprises a system of events. Traveling between home and the office, between departments within the office, and the trips back and forth between a workstation and storage bin comprise a system, a network of those events and those trips. And what happens at the end of each of these trips? An interaction occurs between you and someone else, or an interaction between you and something else.

Open systems, closed systems

Systems can also be regarded or analyzed as closed or open. Decision makers can make a fatal error in judgment at this point when considering including RFID technology into their ongoing operations.

A closed system is "a system whose behavior is entirely explainable from within, a system without input."[6] Why is this important? Because if your organization under study—a manufacturing firm, or distribution center, or warehouse as a system—can be classified as a closed system, then your decision to implement RFID may be a little easier. Use simpler metrics, easier than if it is an open system. A simple example of a closed system is you. From an organizational perspective, you are closed; you are autonomous, and your organizational pattern has largely been determined by your DNA structure. The little circular robot that runs around my house—I have three of them—is a closed system. I push a button and its keeps out of my way, but is persistent a few times, wanting to clean under my feet; its programming allows it to give up, but it seems to remember and eventually tries to come back to that spot.

How many organizations are closed? Can you determine if your organization is a closed entity? When we say systems are closed to information, they can be described as independent.[6] Variations of actions and events outside the boundaries of an independent system do not affect the variation of events or actions inside the system.[6] From a mathematical viewpoint, we say that two variables, M and N, are independent if their values come from chance, not from each other; there is zero correlation between the two variables.[6] The variable M might represent the amount of waste metal in a manufacturing plant that produces metal signs. The variable N might be the amount of A4 paper that is used in the office copier and fax machine for that same manufacturing plant. That specific A4 paper description has no correlation with a pile of waste metal. So, is such a plant a closed system?

Another example of an independent system is your watch or the clock on your wall or desk. Your clock is designed to be independent of the variations of other variables around it, such as temperature, its location on the planet Earth, the gravity of Earth, or the change in gravity as you fly at 30,000 feet to another meeting. However, your watch is not correlated to other watches and clocks, but only to you, the observer. As you walk through the airport, you notice your watch time shows 3:15, and you notice the clock time in the airport hallway is 3:17. Only you see the correlation. Your watch and that clock do not see or sense each other. One does not correct the other's output.

Independent systems, then, are a source of information or a source of data. Independent systems are not the receiver of data or information or communications of any kind.[6] A greenhouse is also a closed system; sunlight gets in, the plants in the greenhouse have some water source, soil, fertilizer, and a human or machine to tend to temperature settings, if necessary. Green plants give off some useful gas in exchange for a chemical process that again is self-contained.

There is one aspect of all these closed systems you may have noticed. That is the boundary: the four walls, floor, and ceiling. However, some energy or work or communication is allowed to enter this boundary. We consider the boundary of the building, the organization, to be permeable. You will have water flowing into it, sewage exiting it, electricity available for operations, doors for goods and people to enter and leave, and doors for materials, raw or processed, to enter and later for some form of that product to exit. Sound confusing? Well, it is, unless you are a thermodynamic systems student or professor or engineer. We will return to the concept of closed systems later as we identify the metrical problems with variables entering the permeable walls or boundaries of each business system we explore.

Closed systems are something we talk about because many approaches to the use of RFID tend to regard the warehouse or organizational structure that is being implemented with RFID as being discussed as though they exhibit closed-system behavior. In the early, what we refer to as classical, methods of analyzing an organization's labor force through time and motion studies, the experts and the decision makers considered their factory as a closed system. The view was, "The closed system organization produced, as efficiently as possible, a standard product or service, which, in the eyes of the workforce, somehow disappeared outside the boundaries of the workplace and was replaced, equally mysteriously, with new raw materials and wages for the workers."[5] Even today, many manufacturing jobs still seem like those at the turn of the century and into the 1960s. People were treated like just another part in a machine and expected to produce their task or widget in a set number per hour with statistically predictable error rates and profits as their metrics.

We still see examples of this closed-system viewpoint in RFID pilot tests that do not seem to succeed beyond that pilot. The reason is the same as the reasons for closed-system mentality having failed in the past: Your RFID system may not be a closed system. It may, indeed, be open. You need to consider if your organization is sufficiently independent of factors and variables from outside its four walls, its boundary, to the world's environment. You need to answer the question, Can you analyze your organization separately?[5]

Current literature discussed closed-loop systems in the manufacturing world, one part of the supply chain. The materials requirements plan (MRP) is used to plan the flow of raw materials, processed materials, and purchases used to make some final product at the end of the manufacturing process. The MRP tells you that a certain task or job has to be scheduled at a certain time of the day and a certain day of the week.[7] Today, operations management of a manufacturing organization is using the term closed-loop MRP. This closed-loop MRP implies that the information feedforward and feedback system between scheduling and inventory control systems is closed. "Specifically, a closed-loop MRP system provides information to the capacity plan, master production schedule, and ultimately to the production plan."[7] It seems that the literature and marketing describe "all commercial MRP systems" as closed-loop.[7]

An open system is one "that changes its behavior in response to conditions outside its boundaries."[6] It was first proposed in the 1940s, and from the fast-paced adoption of passive RFID technology, it appears to be an important subject today.[5] RFID applications, while listed as a pilot, carry a possible fatal flaw of scalability within the whole business process of an organization. The analysis tools and the mathematical models, and the growing abundance of computer models and simulations that anyone can use, should cause the decision maker to be a little alarmed. The reductionist methods that seem to work so well when your car engine fails to start do not apply, and never applied, in the world of information technology, in the computer age, and most importantly in this RFID age.

Today's businesses like FedEx, McDonald's, CRC Press, John Deere, Heineken, and other global enterprises can be considered open systems. A company like Matanuska Maid Dairy in Alaska is a closed system. Fifty percent of its raw material, milk, and other chemicals do come from outside of Alaska. The water, labor, and energy come from within Alaska. The total production, distribution, sales, and waste products are totally consumed inside the state of Alaska's boundaries. Matanuska Maid milk cannot be sold or bought outside the borders of Alaska. (And it has the best chocolate milk in the world, in my opinion!)

However, we will have to consider a new meaning of closed and open systems due to the global environment of today. As we explore the RFID metrics of a global enterprise, we may find a new metric crossover from those metrics used in a closed system versus an open system; we may find that a closed-system metric can now be applied to today's open systems, because today's open systems may be today's new closed systems. This is not just a matter for discussion by academics studying systems or cybernetics. This is an issue for you, the decision maker. Your managerial science lesson is expanding in ways that it expanded in the 1950s. With raw material markets locked in binding contracts from the United States to manufacturing plants in China, and distribution from Finland, for customers in over 40 countries, the vision may be that the open system may be a larger closed system.[5]

Assemblage, pattern, and purpose are the three pillars of our understanding of the metrics needed for the emergent behavior of a supply chain management system based on RFID. We will now examine relationships of this internet of things writ large by RFID.

Decision making

Computers were people who made calculations before 1946, when the first electronic computing machine was made.[3] Since then, the computer machine, along with the decades of manual calculator machines and punch card tabulators, along with paper punch cards and paper tape for storage of data, have been transforming decision making. Decision making before the computer machine was more an art than a science. Along with the computer, and with the birth of operations research and systems analysis from World War II, the decision maker has moved from art to science.[8] However, my experience indicates there is still much black art to decision making left in this computer age, this knowledge age, this Internet age, this age of the internet of things.

Forrester said in 1961, "Any worthwhile human endeavor emerges first as an art. We succeed before we understand why."[8] Since the 1960s, I have witnessed the transformation of decision making using a variety of tools, including, from the 1960s to 1980s, using operations research analysts to ply computer tools to produce some useful number.

The decision maker wants a useful number, a metric. The decision maker chooses alternative actions to pursue, whether to go to war with Iraq or not, whether to purchase the YAH-64 helicopter versus continue with the Huey, and whether to install computers to manage the inventory in 10 warehouses spread around the country. The decision maker chooses a number. That number can be the investment cost of the alternative action in U.S. dollars, EU, or Yuan. That cost is not an isolated number. That same alternative action may have a measure of effectiveness or measure of performance associated with that investment cost.

The complexity of a system's variety

Stafford Beer states, "The variety of anything is its number of distinguish-able elements."[5] By merely counting the number of elements, we can develop a measure of the system's complexity. For example, a light switch has two distinct states of operation. The switch can be on or off. When the light switch is on, something happens; usually a light goes on in a darkened room. When the light switch is turned to the off position, the light in the room goes off; the room is dark. We say that this light switch has a variety of two. That is, the light switch can exist in only two states. This can be proven by many different people or robots or automatic computer signals. As a science experiment, it seems to pass the test of independent testing and proof of its two states of operation. The same is true for the lightbulb. It has two states, just like the light switch. The lightbulb is either on or it is off, corresponding to a parallel operation of the light switch. There is a causal link between the switch and the lightbulb.

What is the variety of the bar code? Like the lightbulb, it can be either on or off. The switch is the bar code reader. Either the reader can scan the bar code as a box passes it along an assembly line, or the code can be read at a retail store checkout counter. What about a passive RFID tag? It has to be interrogated by the radio frequency of a reader's antenna, and the tag will "light up" and send its coded message into the reader's computer. How much variety do a bar code and a passive RFID tag have? The variety of the bar code and of the passive tag and of the light switch are all the same: two. Thus, we define a new metric for RFID, the measure of variety.

The 10 items in the first part of Figure 2.1 represent a collection of similar bar coded items, such as toothpaste. Each small circle is listed with the same number to indicate that each package of toothpaste contains the identical item. We cannot tell the individual classification of each of the 10 items. The variety of each item in this set is still two, if you consider that the number 1 represents the bar code number for toothpaste. Now, that variety is two, dependent on the ability of a switch, the bar code reader, to be able to read the number on each package of toothpaste. The variety in this very small population sample size of 10 items is, however, one. There are 10 distinct

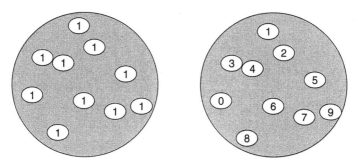

Figure 2.1 Variety metric of a collection of bar codes (left) compared to RFID tags (right).

toothpaste packages, each with the identical bar code number 1, each sharing similar characteristics of definition, a toothpaste package.

Examining the same 10 toothpaste packages on the right side of Figure 2.1 indicates that there are 10 distinct numbers for each item. The variety for this population of items is, then, 10. These represent the use of RFID tags to promote a unique identifier for each package of toothpaste as it exists in the labeling process in manufacturing.

Both diagrams give a different metric of the variety within bar codes and RFID, given the assumption that RFID can be used for unique product identification. These two diagrams present, then, a measure of the information of this small world, or population of packages in some part of a supply chain.

From Figure 2.1, we can readily recognize a coincident of the assemblage of similar packaged bar codes with the identical number. The metric of the inventory is simply 10, but of one item, toothpaste. The assemblage of the 10 RFID items is another story. They also share a metric of inventory that is 10, of one item, toothpaste; but each toothpaste can be identified uniquely, which may further add a new metric of age or date of delivery or date of production. If we make the assumption that the RFID number represents an order in the manufacturing process, a born-on date, then the RFID packages can be said to be linked. This RFID assemblage is then indicative of a relationship among its members, its elements, and its different packages. Because of the unique, but linked, nature of the RFID identification, the 10 RFID packages can be recognized as more related than those with the same bar-coded designation. It is now, at this point, when we realize the RFID items are related, that we also can describe them as comprising a system. The bar-coded items are also a system. Their definition of relationship to each other, however, is not as significant as with the RFID.

The assemblage of RFID items shown in Figure 2.2, on the left, is the information that links the assembly date of each item, showing that item 1 was produced, then item 2, followed by item 3, until the last two items, item 9 followed by item 0. By examining the new information of the manufacturing date, the variety has shifted from 10 separate toothpaste packages to

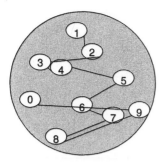 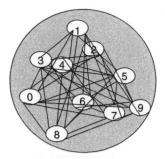

Figure 2.2 An RFID metrical view of an assemblage of the collection of RFID items.

nine sets of relations. But life is rarely linear in operation, so the actual relationship would be more accurately depicted in the right diagram of Figure 2.2. This shows the possible combinations of relationships of each two items in the population. Each RFID package is linked to nine other RFID packages. Because there are 10 packages, there are 90 possible connections. But when item 3 is linked to item 4, there is no need to relink item 4 to item 3. The equation for identifying the number of unique links in this simple population, this small system, is then

Number of connections = $N(N-1) / 2 = 10(10-1) / 2 = 10(9) / 2 = 90 / 2 = 45.$

This means that the variety metric of an assemblage of RFID package population of 10 packages of toothpaste is 45. The unit of this metric is still a simple count. However, the variety metric is now transformed into a measure of information about the RFID population, because we know more about the relationship of the assemblage of toothpaste packages. There is a lesser amount of uncertainty in the information about the 10 RFID items than about the 10 bar-coded items. We have now indicated another possible, yet not as commonly known, difference between the bar code and the passive RFID tags: that of variety, and that of the amount of uncertainty in the use of both.

We have constructed a system. We have shown an assemblage of items, some consistently defined pattern of relationships among the items is known, and now we have some purpose for the existence of this system. We have shown a variety metric that can be calculated to give a measure of the system's complexity, the different number of items in the system, which is 10, but with a possible variety of 45. Obviously, there are many different modalities of product categorization. Some retail stores can stock as many as 100,000 items, which may include 30 different brands of toothpaste. The modality of brand, of different items, creates a scenario where the variety metric of such a system exhibits characteristics of complex systems and complex behavior.

Now, let us take a look at a supply chain, as shown in Figure 2.3. This shows the manufacturing of toothpaste packages, which are shipped

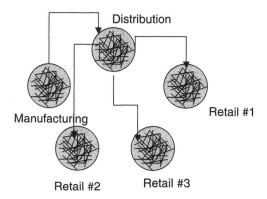

Figure 2.3 Notional supply chain showing complexity and varieties of events inside each node.

to a distribution center or warehouse; the toothpaste packages are then shipped once more to one of three retail stores for sale to customers. The manufacturer makes thousands of these toothpaste packages, which are driven by truck to a warehouse, a distribution center. There, they are inventoried and stored for some specified time period. At some specified time period from the three retail stores, the distribution center will pick up and ship by truck a number of toothpaste packages to the stores. The variety metric in this supply chain model is four, representing the four connections of the assemblage of organizations that comprise the supply chain. You can see by now that when we examine supply chains in more detail in later chapters, the variety of this system will be more than four; the complexity metric will also rise to a higher level, or number, than four.

Emerging complexity

The world saw the development of radar, rockets, and the atomic bomb during World War II. Afterward, the ENIAC computer was developed at the University of Pennsylvania and was used to calculate ballistic tables for the U.S. Army. People in the 1940s and 1950s thought computers were only for scientific purposes, not business. Computers were impractical as a product. In fact, like many, I was hired as a mathematician to program the early computers. In the 1950s, however, the UNIVAC in the United States and the LEO in England started changing the way computers were being used. By the start of the 1960s, when I first began my work on computer systems, there were more than 10,000 computers in operation in the United States. But there were unforeseen problems in this technological change yet to be felt in the government or industry.

Now, enter the seemingly newly discovered radio frequency identification, this seemingly ubiquitous, omnipresent RFID tag, to replace, maybe, the equally ubiquitous, omnipresent bar code tag. Although RFID also had its start in World War II, only in the last few years have miniaturization and lower cost brought it to the mandate platform of change pushed by the DoD and Wal-Mart.

What the early computers did to office management, RFID is doing to transportation time management. What the bar code did to manual price tags, RFID does to bar codes. However, computers evolved from market-driven forces. Computers helped create automated databases and faster flow of routinely processed data. Computers did become a disruptive technology for business. The world of RFID, however, is mandated, driving the market.

To measure the use of a bar code, you need to count; to measure an RFID code, you need to think. RFID will feed the new data stream into the same computer system used by the organization's bar code inventory and tracking system. But the volume could be billions and billions more than the computer systems have ever seen and management has ever dealt with. Therefore, RFID now is becoming a disruptive technology for business. Another direct comparison between bar code and passive RFID tags could be the volume of data that will become the business norm for both technologies.

There is a black box effect in this changing technological impact on today's global supply chains. This black box, which may or may not be a closed system, has a description based on the principles of complexity.

Now, any size organization has complexity. RFID supply chains, or more properly supply nests and networks, are exhibiting increased uncertainties and potential bottlenecks in data filtering. Given this complexity, this uncertainty, is it possible for managers to know all that happens inside the organization, now that it is also global, or how work and product flow really gets done? But not all complex systems are the same. There are some fundamental rules that, when disobeyed, lead to system instability, a failure to learn, a failure to adapt, a failure to evolve, or an explosion.

The issue in the 1970s was trying to find a way to get a close hold on the idea of what was going to happen to the supply chain in 6 months, rather than the next 3 to 6 days. This is still happening today. The old question remains: You may know all about yesterday, but you have to be fairly ingenious to say the right things about tomorrow or next week's supply chain.

The result of this emerging complexity with RFID technological change is that logistics systems with RFID may fail due to a lack of visibility in managing the RFID system's variety. Simple variety attenuators can be constructed to filter and control this variety. This book is a continuing discussion into the studies of large, complex logistics RFID supply nests and network systems. We may not develop a generalized operational theory in this book, but we can reveal the underlying architecture of the mechanics of the system working against or for the logistics technology mandates of DoD and Wal-Mart.

Decision making and industrial dynamics

Decision making, at whatever level of management, "is the task of *designing* and *controlling* an industrial system."[8] The RFID task in today's supply chains is still to design and control an industrial system. The process of that system is still manufacturing some product for a customer: a tea service for a lady in England, or a warm hat for a soldier far from home. Forrester defined the term *industrial dynamics* as "a method of systems analysis for management."[8] This definition describes how we deal with the interactions of all the elements of a system as part of the decision-making process, as part of the tools that the decision maker needs to make decisions on whether or not to implement the newest evolution in computer machine technology: the cheap, ubiquitous passive RFID tag and the active RFID tag.

The task of management has not become simpler since the 1960s. It has become more challenging and complex. Decision makers today, as vice presidents of information systems in a bank, or as operations director in a trucking company, use tools of Six Sigma and total quality management, or hire systems analyst with Ph.D.s in operations research or engineering management. They use tools called neural networks to forecast trends in the assemblage of time-varying data streams of the stock market, or flows of bottled water moving from France to the United States. These tools are still the industrial dynamics of the 1960s.[8] The profession of the decision maker, then, has not been made simpler, but more complex.

Which tool to use can be confusing, as the tools have emerged from the high priest of operations researcher to the decision maker, who has an M.B.A. or M.S. or Ph.D. in management science. But it seems the decisions being made today are similar to those of the 1940s, 50s, 60s, 70s, and 80s, of how much capital to invest in the emerging life of the computer as machine and its data gathering tentacles, such as cell phones, bar code readers, voice recognition systems, and RFID readers. The assemblage of data manuals to optical to radio frequency collection systems is adding a new challenge in the realm of being able to make a decision to redesign your supply chain or logistics operations or transportation system. It is a challenge on how you control those systems, which seemed under control prior to the rush toward cheap passive RFID.

The pace of decision making seems to have gotten faster as the computer speed has gotten faster. That is, the alternative choices a decision maker has to make today versus 10, 20, 30, 40, 50, or 60 years ago is increased. Communications devices have also increased in volume and capability. Computer processing of data has increased in number of calculations per second. RFID is just adding to that volume of data. The decision, however, with more data, needs to be balanced with the decision-making pace of the company.

It is interesting to revisit the definition of industrial dynamics in light of today's supply chain management decisions for companies like Wal-Mart, Cisco Systems, Whole Foods Market, FedEx, and UPS, not to mention the U.S. Department of Defense.

Industrial dynamics was described as a study. Today's decision makers do need accurate and believable data and information to make correct choices, especially in light of the wide range of costs for implementing these cheap passive RFID tags. There are many companies that seem to be studying the RFID investment decision on a 3- to 5-year to a 10-year planning cycle, complete with small pilot tests in selected subsystems of the larger corporate system.

Industrial dynamics, then, is defined as "the study of the information-feedback characteristics of industrial activity to show how organizational structure, amplification (in policies), and time delays (in decisions and action) interact to influence the success of the enterprise."[8] Throughout this book, we will address the quantitative and qualitative aspects of the terms in this definition, such as information-feedback, structure, application, time, success, and most important, interaction.

Identifying the problem

The first issue in examining our RFID supply chain system is to identify the problem. Usually, I hear an "aha" after someone has read about RFID, visited a conference on RFID, or just left the conference where their customer downstream in a supply chain mandated them to now start using RFID or pay a fine for all loads without RFID on each pallet.

Almost all of the problems that have come to me to solve over the last 4 decades have been wrong. The real problem lay buried under symptoms or politics. The RFID technology is not a silver bullet; I do not believe it is even a copper bullet, in the information technology language. RFID is one more root in a system of roots in that decision-making tool called the computer. It sounds easy enough to do: identify the problem and find a solution that fits the problem. However, for those of us who have earned a Ph.D., we all bear the scars of finding that problem to solve years after the dissertation begins. There are many mathematics textbooks and modeling and simulations courses leading to a Ph.D., but you do not need that if you really pay attention to some basic steps.[9] We have some exercises in this book that will challenge you and reinforce your understanding of the problem. These steps have been formulated over decades of education, experience, trial and error, and politics; more on that at the end of this book. If you are on a plane somewhere, read this section and stop for a while. Try out some of the assignments at the end of the book, and do the case study.

Listen to the customer

The first step is listening to the customer. Is your company being mandated to use RFID by some large customer downstream in your supply chain? If this customer is a significant percentage (say, more than 10%) of your revenue, you may have a problem. But what exactly is the problem?

Decision makers are always struggling with customers and customer service. There are internal customers, those working in your manufacturing plant, the warehouse, the retail stores, the stockroom, the distribution centers, the billing office, the freight handlers, the cross-dock supervisor, and the newer and older employees. There are the external customers, such as the buyers of your goods and services. These buyers are either a single customer buying a single item, or a distribution company contracted for quantity, quality, and timeliness of delivery of your goods. Are you thinking of using RFID to improve service to one of these customers? Or are you thinking of using RFID because you perceive a cost savings? Remember, in the system of logistics or supply chain management, the customer is the most important element or component. One of the most important metrics in today's supply chain systems, to today's decision maker, is customer service.

The exterior customer service metric is complex and built around five components: dependability, order cycle time, convenience of order accessibility, communications of tracking and tracing products, and honesty between the company and the customer.[10] There are both regional and global external customers; what appeals to one regional customer may not appeal to a global customer. There are consequences of having poor customer service, such as product stock-out, the order cycle time being too long for customers to wait for the product to arrive, or simply losing a customer. The cost of losing a customer can be as much as 8 times that of retaining a happy customer.[10]

The internal customer demands could also be a problem if management does not listen to them in a meaningful manner. What if management decides it can save money or time (a key metric in business) by implementing passive RFID tags on the windshields of all cars and trucks being delivered to a port from a ship? If that decision is made without regard to the views of the unionized workforce that offloads those vehicles, the shipper, the carrier, and the port could find themselves in an overlooked problem.

So is the customer, who may be a line worker, a retail owner, or a purchaser, asking you to use RFID? If so, that is not necessarily the problem. Experience tells me it is not. That is a good assumption to make, when the customer approaches you. The customers will range from your employees, to the board of directors, stockholders, vendors, suppliers, state and federal government, and international partners and governments, as well as the shopper in the retail store who selects your product to purchase.

Define the problem

The second step is defining the problem. In many of my texts on operations research, this was the first step, but experience teaches me that now this is the second step. But this is the most important step. Miss this step and your decision, and job, are vulnerable to attack. Most problems come to you as a general and often vague idea or notion or suggestion. Hopefully, it is not something as clear as a statement like, "I want RFID tags on all products from now on." You have to see beyond this request or any mandate to use RFID, because "it is not enough for a technology to simply improve the way things are done."[11] Developing a problem statement is the first step that will stop your project if not well thought through.

One problem might be specified as, "Wal-Mart has mandated that our pallets must have passive RFID tags within 6 months or else." The problem is what? This statement as it reads is not the problem statement for your company, which makes potato chips. For the decision maker, the problem has many other dimensions than your customer, Wal-Mart, mandating you place passive RFID tags on a pallet of boxes that are shrink-wrapped. If the decision maker has decided to comply with such a mandate to use RFID technology in a specified manner, your next decision is "How?" The dimensions of such a decision could be the following: identifying the objectives in implementing a technology change program; identifying the assumptions for your complying with this customer demand; identifying the vulnerabilities to these assumptions; defining a rough road map of signposts and business and technology signals or events that could help or hurt such a technology change program; using a systems perspective to analyze how the other departments of your business will be affected by this change in packaging and loading pallets for shipments; again, with a systems perspective, examining a rough approximation of the impact of your adding such tags to your pallets and the carrier or transportation system charged with delivery of your pallets of goods to Wal-Mart distribution centers; and conducting a rough ABC analysis of your customer base.

All customers do not seek the same solutions. Your company's most important customer, say Wal-Mart, might be in group A, producing 20% of the revenue, with group B producing 60% of your revenue, and group C producing the lower 20% of your revenue.[10] Group A might demand compliance with RFID tagging; groups B and C might not think much about it, because they represent maybe several hundred different small to medium-sized retail stores.

The definition of the problem for RFID use is different from many that systems analysts face. Usually, problems are similar to projects, such as a new warehouse. A project uses principles of project management. The project is unique, and you have to build something on time, on cost, on schedule, and meeting the quality requirements. The RFID project is a little different. This project is similar to installing an enterprise resource planning software system that connects all aspects of a company from warehouse, to accounting, to purchasing. To consider that the RFID problem only affects a portion of the organization is too specific, given the systems view of how to assemble and examine the RFID project for new patterns that go beyond a separate warehouse or distribution center. The issue here is to try to avoid optimizing the benefits of RFID for a single entity or single warehouse or only part of the organization. Even with a phased approach, the objectives of the problem should be broader.

Defining the objectives of the decision maker for use of RFID is essential. Samples of objectives might be maintaining profits, increasing the value of the product line, or increasing the organization's position as a bargaining tool.

Assemble the data

The third step is data gathering. To plan for a new technology insertion, you need some parallel data to analyze. There is no data on how the new RFID system will perform. So, you can use past experience in installing a conventional information technology upgrade to help identify possible data needs; also, possible problem areas or issues, which should feed back into the step to define the problem. If you have a pilot program, then this will involve deciding what type of data to collect, how many data fields to record per event, the definition of an event, the method of recording data entry, the number of data entry points, the timeframe to collect a time series of data points, and who will be in charge of this data mining operation. Also, you will need some procedure to validate and verify the data recorded. RFID data collection technology allows data recording of a product every second of the day, 24 hours per day, 7 days a week. Current decision-making policy on materials management, manufacturing, production scheduling, warehousing, purchasing, inventory control, and transportation all become part of the new metric of deciding what data to collect. This part of identifying the problem is where the decision maker has to make some different decisions regarding the flow of information throughout the organization and beyond its boundaries into the vendor's and customer's worlds. Before the plan to collect data is decided, these questions need to be addressed. Again, this will impact the problem and circle back to impact the objectives of the company to employ RFID.

At this point, some analyst will advise the purchase of a new computer database management or information technology system to collect and store and manipulate the data to come from an RFID pilot or implementation. This could be a valid point. After you decide what traditional metrics you normally collect within the current information technology system, you have to look to any changes in the new metrics or units of current metrics being collected.

The data collected should first come from the current process of the operational supply chain. This forms the basis of comparison of new data that could be collected through an RFID system.

Beyond the data collection is the data assemblage of this data. If you are a small to medium-sized company, you may not have a staff of operations researchers or statisticians or economists on hand to help you find patterns in the data. Collecting the data and formulating a view of the data are linked; when the pattern is recognized or shifts, the data collection process may also change.

Formulate a view

The fourth step is formulating some kind of view of the problem as a model, diagram, mental, causal, on paper or with a computer. This is where some pattern within the data will emerge. And there will be a pattern. It may appear chaotic, random, or with some regularity. It must be tested statistically to satisfy findings if the data assembled has specific meaning as expected or as a surprise to the decision maker. From this assemblage of data, we will obtain a purpose of the data, of the system under observation.

A simple statistical analysis of data that represents the movement of items or goods between two points is a good starting point. This is probably already being collected over some time series. Basic statistical descriptions should be run on the data to at least look at the mean metrics and standard deviations around that mean. There will be more basic statistical analysis described later. The metric at first may be as simple as customer wait time (CWT) or requisition wait time (RWT). This would be useful data to apply to the metric of customer satisfaction based on order cycle time. The metric may be the number of days between processing points, such as shipment of goods from a manufacturer in China, to a port in Alaska, to a distribution center in Memphis, on to a customer in Europe.

The view of the data stream and its movement is essential to allow the human to observe, in an abstract manner, how the company is doing business, and how the lifeblood of the system is flowing.

If you are using data from another, similar company, from some of the sources cited in this book, for instance, do you know when the data was collected, why, and how it was or has been used by others? When you look at how other organizations have implemented RFID systems, be careful of transforming their wonderful experience directly into your expectations. All data is not alike. Seeing and playing with the data is a critical step. Many will also try to leap from cause to effect with the first pattern they see in the data. Rarely are data trends or patterns so obvious that you can make a new discovery from them. At best, you get a feel for what the data is telling you. It is still your job or the job of someone with enough statistical knowledge to find out a little more about that data's story. We will give you enough of such statistical analysis to get started along this path.

Develop a model

The fifth step is creating a mathematical representation of this model, based on the level of your skills in mathematics. This is where many people get off track and lose sight of the system they so carefully articulated. If you are a small to medium-sized company, you may want to invest in some consultant time with the local academic institution in your area. When developing a model, reality is copied. But that reality of the real supply chain is just that—a copy—and not very real. There is no model that depicts the entire supply chain exactly as the real system operates each second of the day. The delays in process, functions and logistics threads—arcs and nodes that connect parts of subsystems and the overall system—and the unknown unknowns that infect human-based systems cannot be totally modeled. We keep trying. But we do not have that model today.

The model that is developed will possibly be a closed system. That is a simple way to show the connection and interactions of all the parts of the system.

A first step in the model building should be a group discussion of all the interested stakeholders, the decision makers, managers, analysts, and workers in your organization who may be owners of the problem or part of the problem, and will be impacted by the solution. This aspect of the problem definition is critical. It is explored in more detail in Chapter 6, on assumption-based planning.

The result of this step is a model. But this model may be a wall full of diagrams and charts and arcs and arrows showing the flow of the process leading to uncovering the problem and possible solutions. Because this entire book is about RFID, the process, the model will be about the implementation question of RFID. Therefore, the model will have a central theme of some alternative RFID implementation decision action at its center or focus. Also, there should be a thought process, drawn out graphically, that shows how the business process interacts with the RFID technology and the analysis functions. That is, the model should show how analysis, technology, and process integrate to form an implementation solution or set of solutions. This model will more than likely be what we call a mental model, a causal diagram, or a flowchart. Many times, the artifact of such sessions is a wall full of white paper or series of whiteboards with the mental model outlined for discussion, along with a list of key factors that define each node and each arc connecting each node.

Designing a first cut, mental model at this stage can sometimes be all that is needed to give the decision maker the visual tool needed to decide to proceed with RFID or not. This process is similar to what you do when you are faced with 10,000 requisition records and you want to plot them over time to see what pattern is emerging before you spend time on more detailed, computer-driven, statistical analysis. This visual, feedback-linked diagram will indicate the boundaries of your organization as a system. You need to see and believe, with factual data for support, whether your organizational process is a closed system or open system. This is not an easy task.

If you are convinced or have the capability to take your mental model into computer graphics with its dynamic, then you have a better chance of playing with the factors and dimensions over time to see what impact RFID technology will have on your process and how best to continue to analyze the key metrics. The time factor is really only introduced if you can make a visual computer model of the process. This is not as easy as the software providers make it sound in their marketing brochures. How to avoid some of these pitfalls will be detailed later in Chapter 7, on modeling and on forecasting.

Exercise the model

The next step is creating a method of exercising this model. This can be with computer models or simulation building or qualitative assessments, if you cannot afford such talent. The exercise is where you will examine the data that is used in your model. Models are simply algorithms. Like all computer-based algorithms, these algorithms will calculate whatever you program them to do. Consider, for instance, a situation where you have a number of 12,000 order requisitions and you divide that number, by mistake, by the price of a half-gallon of milk costing $3.00. The result will be a number, 4000, but what is the unit? If you follow mathematical principles, then the unit will be orders per dollar, or 4000 requisition orders per dollar spent on a half-gallon of milk. This illustration is part of the problem for people who have been developing models for just a few years. Anyone who is computer literate can download a model-building set of software, but the complexity of depicting the actual process with the correct metrics unit gets lost too many times.

In the conference room where the mental model was diagrammed on the whiteboard, you really have a snapshot, a static image of the process frozen in time. It is not real. It has a visual depiction of the real parts of your organization.

Test the model

Then, test your model. This is where the model is tested with the variables that you choose. Neither the model nor the computer chooses for you. You have to know the metrical units that compose the components of the system you have emulated. Remember always that this is a model; it is not the real thing; for that, just go down to the loading dock and watch how pallets are really loaded using the metrics of the forklift driver, not your diagram made in a clean office away from the noise and movement and complex variety. This model can be tested as a manual exercise. In fact, if you can and the model can be depicted on some projector or whiteboard or papers lining the walls of a conference or classroom, do so. You will find out more about your metrics this way than with the computer simulation of your real-life system. But most systems are too complex and your time is too valuable for this type of manual exercise. So, the computer output must constantly be challenged when you take this model out for a walk. And there are no dumb questions when it comes to asking why some metrics value is coming out too high for your expe-rience. Experience has also shown that your experience is more valuable than that of the model; that is why companies hire high-priced CEOs with lots of experience to run a complex company, which is a complex system with too much variety to see, let alone control. We still have a tendency in many companies to act like Henry Ford and try to control all that variety from our armchairs.

Develop an implementation plan

The next step is to develop an actual implementation scheme. This sounds simple, but it is as complex and time consuming as any of the other steps. The implementation plan is much like that of a plan for a project. If you have project management experience or training, then this would be a big help in formulating a plan. This means paying attention to the tasks to be completed, the schedule of tasks to be completed, the cost of each task, the linkage of each task to each other task, and quality control along the way toward implementation and after implementation. Developing a flowchart or a process chart will be useful and helpful, as all project managers know. So, how do you start? You start at the very end of the plan. What is it you expect this project to produce? What metrics are you going to measure for success? Is it simply more customer orders, or a certain amount of profit or a zero tolerance for inaccurate inventory counts per week, day, and hour? From this point, you then work your way back to the beginning point. Along the way, you keep examining the system components, checking for problems or variety, of data collection, of data that may have to be aggregated from one point to the next.

Feedback

This step is not really a step, but a process called *feedback*. You must plan to revisit all steps. This sounds simple. It is not. Decision makers will not want you to spend too much time on revising or revisiting decisions on metrics and linkages that have already been decided in previous stages. If this is the case, then I would walk away from the project. It is going to spin out of control sooner than later.

References

1. Eiseley, Loren, *The Firmament of Time*, 1st ed., Atheneum, New York, 1960, 17.
2. Drucker, Peter F., *Managing for the Future: The 1990s and Beyond*, 1st ed., Truman Talley Books/Dutton, New York, 1992, 312.
3. Brown, Lesley, (Ed.), *The New Shorter Oxford English Dictionary*, Clarendon Press, Oxford, U.K., 1993.
4. Ljung, Lennart, *System Identification: Theory for the User*, Prentice Hall PTR, Upper Saddle River, NJ, 1999.
5. Beer, Stafford, *Decision and Control: The Meaning of Operational Research and Management Cybernetics*, John Wiley & Sons, Chichester, U.K., 1966.
6. Krippendorff, Klaus, *A Dictionary of Cybernetics*, University of Pennsylvania, Philadelphia, 1986.
7. Arnold, J. R., Chapman, Tony, and Chapman, Stephen N., *Introduction to Materials Management*, 5th ed., Pearson/Prentice Hall, Upper Saddle River, NJ, 2004.

8. Forrester, Jay W., *Industrial Dynamics*, The M.I.T. Press and John Wiley & Sons, Inc., New York, 1961.
9. Hiller, Frederick S. and Liberman, Gerald J., *Introduction to Operations Research*, 7th ed., McGraw-Hill, Boston, 2001.
10. Gourdin, Kent N., *Global Logistics Management: A Competitive Advantage for the New Millennium*, Blackwell Business, Malden, MA, 2001, 41–52.
11. Smith, Cooper, *Technology Strategies*, Prentice Hall PTR, Upper Saddle River, NJ, 2002, 193.

chapter three

Applications of RFID and how to view them for your use

> "...you may choose to challenge the trends, but first you must know where they are headed."[1]
>
> **John Naisbitt and Patricia Aburdene**

Current applications for RFID

There are numerous case studies for the student or reader who wants to try to find an application for using RFID technology. Actually, your imagination is your limiting feature. A few applications are listed here to start you in the right direction. But be aware that of all the applications we have seen, more than 90% of them are suspect, or not well thought out. The problem that you see is usually a localized problem and not a global or strategic problem. And in this global world, the supply chain is more a network of operations, even in one organization.

Applications of RFID are really everywhere you look. First of all, if you own a fairly new car or truck, you have an RFID device in your pocket. Take that car or truck key out of your pocket. If it has a button on it to push to open the door, or lock the door, or set an alarm, that is an RFID. It is a wireless transmitter to your car. You are transmitting a radio signal of low frequency to a radio receiver in your car, which triggers an alarm or a locking or unlocking device.

Applications for RFID tags have been around for years. What is changing, however, are the applications with passive RFID tags to comply with the mandate from Wal-Mart and the Department of Defense for their major suppliers. This is where most of the newspaper and trade magazine articles are concentrating their fears and desires and trying to answer the question of applications to meet this mandate. But while the suppliers are looking for the best-case scenario to meet the current short-term demands, there are plenty of examples of applications of the passive RFID tags that can help provide evidence on applications. There are plenty of examples of transportation smart cards or tickets being used for public transportation such as with the rail systems in Europe (but not so widespread in the United States). However, not all countries in Europe subscribe to the use of such smart tickets.

Access control into public buildings and government buildings is already under way in the United States and has been for some time in Europe. In the United States, especially since 9/11, the use of passive RFID tags in personal identification badges is commonplace. Around Washington, D.C., you see this everywhere. You cannot go into a federal building without at least one or more identification badges, one of which is RFID.

RFID tags have been used for animal identification for years, but they are usually active tags—with a battery. They help identify animals from birth to death and even monitor and identify what and when the animal feeds.

Another security RFID application is in the tracking of humans for the justice departments of the world. We have all seen high-profile criminals "incarcerated" in their homes, wearing active RFID tags, often with GPS sensors to track their locations in real-time.

Perhaps the most noticed and written about use of RFID tags since 9/11 is in the cargo container field. With millions of containers entering the ports of the United States each year, only a few are actually checked for their contents. The use of RFID tags from the point of origin into the port of entry in the United States or Canada is receiving a great deal of attention from the Department of Defense and the Department of Homeland Security. This is one area that still remains to be tested and proper process for use identified.

Sporting events have been using passive RFID tags for some time, as well. Skiers and runners know that passive RFID tags have replaced paper-like tickets.

Industrial application in the automotive field has been a leader in using passive and active RFID tags for some time now in the United States and in Canada. The assembly line is now more just in time than ever before.

RFID tags in the medical or health care field is under scrutiny. There are some pilot tests under way to tag patients in an emergency room, to keep track of them. But there are also privacy and ethics issues that have yet to be solved. Until these ethics questions are answered, the use of RFID for patients may be slow in coming.

Since 9/11, there has been an initiative by the United States to have smart borders with Canada and Mexico. These smart border applications are designed to prevent terrorist attacks within the United States. How companies invest in systems such as RFID to ensure their trucks of goods can cross between the border points of the United States is an issue of costs and revenue. The company can invest in the RFID technology to help the border and customs organizations identify trucks and containers that pose no or little risk to the United States. But this RFID option does cost, and there are options available that do not rely on RFID technology.

Libraries are adopting passive RFID technology for patrons to self-charge their books, CDs, DVDs, movies, or other artifacts. Libraries seem to be experiencing faster times for charging and discharging books and other library materials used by the patrons, faster inventory checking, increased accuracy of inventory, faster shelving of books, and overall faster handling of the library materials, as well as reduced theft of books. The RFID metrics for the library appear to provide more accurate data on the traditional metric count of books. The distance between the RFID tag and reader or sensor seems to be 2 feet or less. The readers, called sensors, are usually placed at the library building exits. These appear to be fixed sensors, mounted usually one on each side of the exit. Two are used for those hallways or exits that have a width of 4 feet. If an exit has a larger width, then RFID tag sensors in the ceiling or floor might be an additional design feature.[2]

One application that will impact all U.S. citizens would be if the U.S. Postal Service used RFID tags. The U.S. Postal Service manages billions of envelopes and packages per year, within the United States and coming and going from outside the United States. One possibility within the U.S. Postal Service is to use passive RFID tags to create a smart stamp. Such a concept could have an unlimited array of uses, but all with hidden costs. Such smart stamps could be coded with more than just the cost of the stamp. It could be used to indicate who is sending the letter and who is to receive the letter. As the letter moves through the postal system, it could be tracked from post office to distribution center to carriers to post offices and more carriers and finally to your mother's mailbox a thousand miles away. You could even have smart mailboxes. With smart mailboxes, the mailbox RFID antenna would instantly notify you that the specific coded letter you are waiting for is there, and the sender would know that it had arrived. This information could all be coded in such a way as to read "sent" over the Internet. You could track the movement of each letter much as you can track your letters today, but with more accurate visibility along the chain of custody. The capability already exists for you to create your own smart stamps at your home PC and with a modified printer. Or you could go to a postal kiosk in a store or in the post office and create your own smart stamp.

The Swedish Post is already using RFID to detect and record parcel tampering. Microsoft Corporation is also helping the U.S. Postal Service with this ubiquitous RFID tracking and tracing service. Using a system called MSN Messenger, the entire route of the smart stamp can be traced from beginning to end. The RFID smart stamp or tag is about the size of a playing card. This RFID tag can be combined either with the cost of the stamp (it would be output on a simple printer), or as a stand-alone package processing tag. It would be interesting to talk to some of the technology change leaders of our time or from our history. For this postal system, Benjamin Franklin comes to mind. As a forward thinker and as the first Postmaster General, I wonder what he would say about all the fuss over use of RFID tags for smart stamps, or for stamps for that matter.

The list of applications for RFID technology is growing almost daily. Routine checks on the Internet uncover many developers ready to provide products and services. Some of these are Microsoft, Symbol Technologies, Wincor Nixdorf, Neopost, GATS, Lyngsoe Systems, Lijnco, CODEplus, WINN Solutions, Denstron, Texas Instruments, KSW Microtec, Identec, Savi Technology, CSC, Trenstar, ID Systems, Ask, NBG ID, Hala Supply Chain Systems, EM Microelectronics, Trimex International, Escort Memory Systems, ECO Co., Intermec, UPM Rafsec, SamSys, Hi-G-Tex, NCR, Avonwood, Cypak, Avery Dennison, Alien Technology, FKI Logistex, and Baumer Ident.

Figure 3.1 shows a snapshot of applications from around the world that anyone can find from researching the nearly 200 countries on planet Earth. Two areas are the primary focus: China and the Russian Far East. China was chosen due to its emerging dominance in the supply chain and RFID business environment; the Russian Far East was chosen to show a developing part of a developed country that is just now entering the use of RFID for retail.

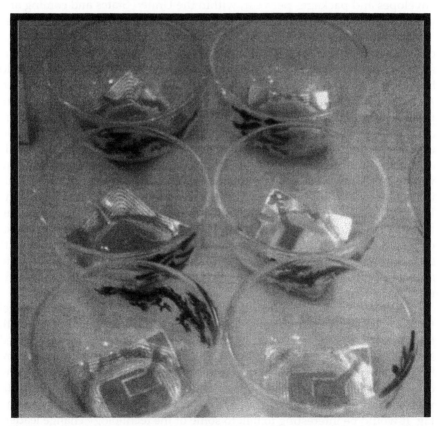

Figure 3.1 Chinese beverage glasses, sold in Parus, a Vladivostok, Russia, retail store. Photo taken in April 2006 (photo courtesy of Lyndsay Miles, research assistant to Dr. Hedgepeth).

RFID adoption in China

The invisible hand of Adam Smith is the obvious hand controlling market forces in China. The majority of Wal-Mart suppliers in China are delaying the implementation of passive RFID tags in compliance with the Wal-Mart mandate. Chinese retailers do not seem to have the same degree of competitive leverage over their vendors as experienced in Western countries, such as experienced in the U.S. Wal-Mart bidding relationship with its vendors. The supply chain system in China is not as developed as it is in Western countries. Chinese businesses and government officials also do not seem to have an understanding of how logistics management and supply chain management operate from the viewpoint of Western systems.

Chinese vendors seem to not be embracing RFID as a mandate from Wal-Mart, as is the case in the United States, or as a mandate from the government, which is the case in the U.S. Department of Defense. There is evidence emerging that information about RFID technology is not known to Chinese customers. RFID seems to be difficult for Chinese businesses to understand. Other evidence indicates that Chinese businesses are concerned with ROI, considering that the cost of RFID may not translate into profits in a short timeframe. Another barrier to total supply chain implementation of RFID is the apparent lack of information technology infrastructure. So the application of RFID appears to be one destined for tracking and tracing goods as they enter and leave a dock door or other storage facility within a warehouse. The metric, then, that appears to be most useful to China for cost of goods (COG) is the inventory count and control of shrinkage or products.

The most important issue for China

The government of China, in Beijing, seems determined to promote the use of RFID tags in selected areas. However, the application of passive RFID for China businesses is significant in selected areas. The Chinese government appears to be the motivating factor behind the use of RFID rather than the manufacturing and distribution businesses. The Chinese Ministry of Public Security seems to be the leading user of passive RFID tags. This ministry is issuing identity cards, much like a driver's license in the United States, that contain personal data, a photo, and a passive RFID tag. There are 1.3 billion people in China; each one will be issued such a smart identification card. Reports from 2005 indicated that over 100 million RFID tags were purchased and imported by the Chinese government. Further reports indicate that the Chinese government is purchasing RFID tags for a range of prices, from 15¢ to $200 (U.S.). This would indicate that the government is pursuing not only passive, but also active, tag applications. The goal is to have all 1.3 billion people as RFID consumers by 2009, by the issue of simple personal identification cards.

The following cities or provinces have ongoing programs with the use of RFID: Shanghai municipal government, Shenzhen, Wuhai, Beijing, and Haerbin. The different ministries involved in the application of RFID appear to be the Ministry of Public Security, the Ministry of Shanghai Port, and the Ministry of Chinese Science and Technology.

In Shanghai, the government has backed the use of passive RFID tags on farm produce in part of the Shanghai Wholesale Center. Reports indicate that there is a passive RFID tag on each item of produce: potatoes, apples, and so on. This appears to be a pilot study in the use of passive tags, since the Shanghai Wholesale Center sells and distributes only 20% of the entire produce market of Shanghai. This same wholesale center experimented with the use of bar codes on produce in 2001. However, bar codes in this scenario appear to have failed. The bar code problem appears to be one of the paper tags not adhering to the produce. This is again a government-subsidized venture; the cost of a 15¢ passive RFID tag would be prohibitive for produce that could cost the same or less than 15¢ per item.

China is using passive RFID tags on pets for identification. We are making an assumption that this program is also being underwritten by the government of Shanghai. Shanghai is also using passive and possibly active RFID tags on hazardous waste transportation.

When you read between the lines from the hundreds of articles about RFID use in China, you find that the Shanghai government is trying to leap forward in improving parts of its supply chain, although those supply chains are not as integrated as those in Western countries. Besides tracking people and pets, and experimenting with produce, the most significant application will be in the ports. The Shanghai government has contracted with the Shanghai Internationals Port Group, Ltd. (SIPG). This port experiment is aimed at using this RFID technology to tag and trace the billions of shipping containers that move around the world, many starting from Shanghai. The adoption of RFID for Shanghai port containers appears to be a move of the Ministry of Shanghai Ports, which was privatized in June 2005.

Metrics for use by the Chinese government appear to not conform to those of the Western countries, United States, and Europe. China appears to be implementing RFID technology without regard to quantitative metrics. It appears to be using a qualitative metric of compliance with the Western countries that are using or contemplating RFID. If China tags its cargo containers, then the position of China as a world-class player could be a valuable promotion tool.

Metrics for China's use of RFID are dependent on China's government, industry, and academia understanding how to reform their current logistics management system and supply chain management infrastructure. Along with this understanding of how to reform the infrastructure is the metric of labor. Currently, the Chinese metric of a low-cost labor pool is driving a significant advantage for manufacturing goods in China. A possible metrical issue in the next 10 years will be the cost associated with this labor pool. The RFID tag cost will possibly remain significantly higher than any savings from the Chinese labor pool for many years.

An implementation strategy China may have to consider is using a closed logistics system, rather than an open system. With the government backing RFID implementation, any closed system could possibly show value added to the products flowing though a process. This does not mean it will be cost effective as ROI is calculated in Western countries. The metric for use in such a system would have to show a decrease in shrinkage, improved inventory reporting, time savings in order picking within a warehouse, and accurate data for products delivery. Cost of goods would not be a metric for the China RFID model. Open system architecture, then, is an issue in China.

The China metrics currently are problematic in data collection and processing, communicating the flow of product information throughout a supply chain, and standardization of processes. These problems are not for just one geographic region of China, but for the dispersed factories around China. It is not a guarantee that the business culture in a different province for the same brand manufacturer or vendor will be followed.

For the foreseeable future (10 years), China will need five events to take place from the West: (1) training and education programs for Chinese universities and industrial institutions, from Western academic and industry trainers, in how Western logistics and supply chain management operates; (2) a broad view on how to determine and use RFID technology; (3) an understanding of how globalization is affecting Chinese business ventures across all provinces of China; (4) an understanding of how China can adapt Western metrics systems to a Chinese retail and wholesale industry that seems to be subsidized by the government; (5) and a better understanding of the differences between Chinese business culture and the United States, or between China and the U.K., China and Germany, and so on.

RFID adoption in the Russian Far East and Moscow

There is little evidence to suggest that RFID is being used on a wide scale in Russia, and it is almost unheard of outside the major corporations, Moscow, and St. Petersburg. However, in certain areas of industry and research, knowledge of RFID appears significant. The Russian Far East covers a vast amount of Russia east of the Ural Mountain range. In the port city of Vladivostok, RFID tags began showing up in a retail grocery store in 2006. This grocery store, Parus, opened in May 2005. It is an upscale business, but basic grocery and household items remain competitive with other local supermarkets. In 2006, passive RFID tags started showing up on a few individual grocery items, such as instant coffee and packages of walnuts, both of which are high-priced items and a luxury in this part of Russia. Kitchen items such as glassware have passive RFID tags as part of the bar code tag on the bottoms of each glass. During 2006, Parus began displaying more passive RFID tags on nonperishable items, especially items that were easily stolen. The purpose of these passive RFID tags appears to be security. If an individual item is taken past the antenna near the door, an alarm goes off, indicating a possible theft of that item. The clerk uses the bar codes in a scanner to record the cost of the item; however, the bar code seems unknown to the clerk or the shopkeepers. There seems to be no indication from the store workers as to what information is actually stored on the RFID tags.

There are several news stories published in the Russian press about RFID implementation in Russia. Avtogaz (GAZ) is one of Russia's largest automotive manufacturers and a member of the RusPromAuto holding. In 2006, Avtogaz completed its first passive RFID project. The reports are that the company will have a RusPromAuto-wide application of passive RFID tags between 2008 and 2009. Through the use of passive RFID tags, it appears the company is trying to be a leader in just-in-time service for its customer, a concept that is rare in the Russian Far East.

Throughout Russia, considerable capital investment is being made in passive RFID technology. There are mixed reports on the use of other digital technology applications. One report from Estonia indicates that it is using digital technology for the police and for political voting over the Internet. Estonia appears to embrace technology like RFID, but the reports are sketchy.

One aspect of technology advancements using RFID for Russia is an investment culture. There is a strong and growing investment culture in research and development; however, funds usually go to mature companies, not to start-up companies. Start-up companies and entrepreneurs are not generally supported. One publication conducted an innovation contest in 2004; the winning entry was an RFID application. However, this winner has not been funded even after 2 years.

There are scattered plans for the use of RFID technology for the future. In Moscow, the Metro Cash & Carry stores (part of Metro Group from Germany) are using and plan to expand the use of passive RFID tags for security. This effort in Moscow is to parallel the Metro Group store of the future in Germany.

In Russian agriculture, automation of harvesting and processing of produce is considered the ideal method of farming. There are many press stories on how automation has saved labor and standardized the produce quality and minimized human error in handling produce.

On a rabbit farm in Kazan, Russia, the Miakro Karatau's farm has been using technology for raising and harvesting rabbit meat and fur since 2002. The farm provides its meats and fur to the Metro stores in Moscow. Each rabbit on this farm has a personal "passport." This passport contains a rabbit's history from birth date, weight history, transplant from the farm, and the optimum harvest time for the rabbit. A computer database contains each rabbit's history and production schedule information. Rabbit farmers now use computers as one of their farm machines.

Miakro Karatau is implementing a passive RFID passport for each rabbit. The goal of the farm is to create an autonomous sensory production network, which will monitor all growing requirements of lighting, temperature, humidity, vibration, food, and water for each rabbit, from birth to harvest.

The public libraries in Moscow are investing in RFID technology with tags that are called protective-analytic. The cost to each library is approximately 8.5 million rubles (approximately $303,571 [U.S.] in 2006) per library. This cost is approximately 10% of a library's development budget in Moscow.

Russia is also investing in a biometric passport system. Each new Russian smart passport will contain the addition of a passive RFID tag. The RFID chip will duplicate all the printed information on the passport, plus have a digital image of the passport owner's face.

Alaska wild harvest seafood

The majority of food consumed in the United States originates on a farm. Whether a given farm is within or outside the United States, farms, both agriculture and aquaculture, produce basic foodstuffs, which then enter the general food supply chain on the way to the consumer, as shown in Figure 3.2. The U.S. seafood supply chain is similar to a general food supply chain, in that the products flow toward the end consumer along a chain of custody supply chain that is heavily regulated by federal laws. There is one difference: a supply chain of perishable food products that are farmed, which could be animal, plants, or fish (freshwater or from the sea). Wild Alaskan salmon and other species of wild seafood, on the other hand, are hunted. The distinction of farmed versus hunted food has become significant in the use of time and temperature devices tracking the cool and cold foods through the chain of custody from farm harvest or sea harvest to the white-tablecloth end consumer in a restaurant or home kitchen.

Steve Grabacki, president of Graystar Pacific Seafood, Ltd., said, "The difference between farmed and hunted food makes the supply chain of all wild seafood far more complex than that of other farmed produce, meat, and fish."[3]

In 2003, the federal and state waters off Alaska yielded over 5.3 million pounds of finfish and shellfish, worth approximately $900 million before processing.[3] This cost was paid to the seafood harvesters—the fishermen. It is called the ex-vessel or dockside value. This quantity of wild seafood harvest is more than half of all seafood produced by the United States.[3] And with natural disasters such as Hurricane Katrina, which devastated the wild shrimp industry in Louisiana in 2005, the value of other wild-caught U.S. seafood becomes higher in price and profit.

In contrast to farmed food animals, such as cattle, swine, and poultry, wild fish have a mysterious quality. It seems that wild seafood is almost invisible. The wild seafood swims about in the ocean as it chooses. The seafood metric is simply the number of the species that are caught each year in pounds. To the average consumer, they may see fish as three or four species, salmon, trout, catfish, and tilapia. In fact, there are five species of salmon, then there is pollock, sablefish, many species of flatfish, various forms of crab, halibut, various species of cod, various species of rockfish,

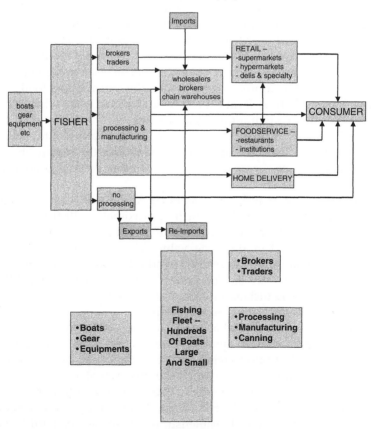

Figure 3.2 Alaska seafood supply chain with possible RFID reading locations.

bluefish, tuna, and the list goes on. When you purchase a fish sandwich at your favorite fast-food restaurant, do you really know from which species listed above the fish fillet was made? The environment of wild seafood is also one that is hostile to humans, because unlike farming, wild seafood is not under human control. Humans cannot make more wild seafood, as they can increase farmed products. This means that the wild seafood population is finite, growing at the control of Mother Nature, with the decrease in population coming from humans.[3]

From a systems and complexity perspective, the wild seafood industry supply chain has at least five sources of uncertainty: technical, biological, spatial, political, and economic.[3] The use of RFID time and temperature tags becomes the dimensions and metrics of the wild seafood chain of custody. The decision to use RFID time and temperature tags rests with many carriers within the supply chain, from the captain of the fishing boat, to the purchasing agent, the processing plant, the truck carriers, the airline carriers, the distribution warehouses, the wholesale and retailer, and the chef at a restaurant.[4]

References

1. Naisbitt, John and Aburdene, Patricia, *Megatrends 2000: 10 New Directions for the 1990s*, Avon, New York, 1990.
2. Boss, Richard W., RFID Technologies for Libraries, Public Library Association, May 14, 2004, http://www.ala.org/ala/pla/plapubs/technotes/rfidtechnology.htm (accessed April 28, 2006).
3. Grabacki, Stephen T., The Supply Chain of Alaska Wild-Harvest Seafood, 2006, unpublished paper.
4. Smith, D. and Sparks, L., Temperature controlled supply chains, in *Food Supply Chain Management*, by Bourlakis, Michael A. and Weightman, Paul W. H. (Eds.), Blackwell Publishing, Cornwall, U.K., 2004, Chapter 12.

chapter four

Traditional metrics for logistics and supply chains

> "Information is data endowed with relevance and purpose. Converting data into information thus requires knowledge. And knowledge, by definition, is specialized."[1]
>
> **Peter F. Drucker**

What is a metric?

Metrics are the linkage between a businesses strategy and its operations.[2] Metrics are also linked with the term *measurement*.[3] The value of understanding what metrics mean to the decision makers has continued to grow with the growth of computer technology since World War II. The current utility of metrics in decision-making tools in the world of supply chain and logistics has also increased in importance since the increase in operations research (OR), led by the historic use of military operations research. There are several professional organizations catering to the discipline of operations research, and from that rich history, this story of the application of new metrics to RFID has emerged.

In the information technology part of the study of supply chain activities, metrics can be seen as a useful management tool for "tracking and evaluating performance, benchmarking, feedback, and organizational improvement."[3] In the materials management part of the study of supply chain activities, metrics can be seen as a tool without which "no firm could expect to function effectively or efficiently on a daily basis."[2] For the materials management decision maker, metrics provide a tool for control, communicating data up and down the management chain, corporate learning, and business improvement.[2] These metrics are critical to examine the expectations of decision makers, stakeholders, and stockholders; help identify critical problems; direct courses of action from manufacturing to transportation; and the usual suspect, motivate people to action.[1]

Why we concentrate so much on metrics is that decision makers are production control professionals, whether that production is based on raw materials consolidation at the beginning of a supply chain or the customer reaching into the seafood section at a retail store and selecting the freshest seafood. RFID use today as part of the tool of metrics is driven by reasons that are still solid in the minds of decision makers. These are, at a minimum, that customers are never satisfied with the product choice, the timing of the delivery, and the quality of the product. The supply chain is no longer regional, but growing larger each year and defying decision makers to understand the rules of the science of complexity. Product life cycles are getting shorter and shorter as the retail supply chains move from a push system to a pull system—except for the military, which remains the hallmark of push supply chains. There is a vast amount of data being created over the Internet, with sophisticated, integrated information technology hardware and software, and now the threat (or blessing) of RFID data production. There is always the profit margin to consider, even if you are a big oil company that seems to make record profits moving that liquid gold, which we buy for many products, not just gas for our cars. There are an increasing number of alternatives to purchase, to plan, to produce, and to transport, all made more visible or transparent along the supply chain by RFID.[2]

There are corporate strategies and corporate assumptions supporting that strategy on how to interact with customers in providing goods and services, whether these customers are internal to the company or external. So, it is the metrics that become the bridge linking this corporate strategy to the operations of the company.[2] Figure 4.1 shows this linkage. The left side shows the inter-action of the corporate strategy, its core assumptions over a set timeframe, and the view of whom and how to treat all manners of customers. The metrics form the bridge between this strategic thinking and the operational implementation of machines and people to operate and interact with machines. Now, these machines may be material manufacturing machines or computers (which are, after all, simply machines). The metric for the use of RFID technology has to go back to the beginning of the corporate assumptions of the strategic thinking about the customer. The metrics are defined here. But this

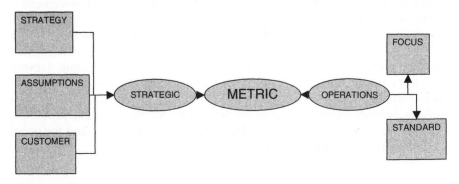

Figure 4.1 Metrics connections of company strategy and operations.

should be balanced with the right side of the visual equation, which deals with how to implement these metrics. The focus is where we define the specific RFID activity to be measured. Within these RFID metrics, we will also need to define standards as a baseline performance against which the RFID metrics will be judged as meeting or not meeting the standard.[2]

The performance measure of some application of RFID has to be turned into some quantifiable number. If the measure is qualitative, there are relative measures that can still convert these qualitative aspects into a number from 0 to 1, and still be useful for analysis. All RFID metrics, then, will be defined, as all materials management performance measures are. The type of number most often used, and the strongest statistically, is the ratio. We will look for ratios such as quantity of inventory; say 10,000 items, over a period of time, such as a week. So, one RFID measure could be 10,000 items per week. Of course, there is a geographic location with each number. Thus, this 10,000 items per week might be in warehouse number 3, with 20,000 items per week as the count in warehouse 1, with warehouse 2 being empty due to construction. So, metrics or RFID measures should consist of at least two parameters.[2]

If your company is deciding to implement RFID technology as a pilot or as a total systems integration, you will need to focus on the performance standards. These are created from a process of "transforming company policies into objectives and specific goals."[2] Each goal will have a target measure for the RFID application.[2] For example; with bar codes, you may have a target of 96% accuracy on the read rate of cargo at the loading dock. However, with RFID you may have a target of 99.9% accurate read rate. As Arnold et al. said, "Performance standards set the goalpost, while performance measures say how close you come."[2]

Today, how to properly use metrics or performance measures is a common practice by the company's operations research analyst. However, there are many small to medium-sized companies that may not understand the complexity and value of performance measures, especially for disruptive technologies such as RFID and (at one time) bar codes.

Performance metrics for implementing RFID technology into current supply chains seem to focus on traditional indicators of time and units. Changes to metrics may happen if the decision-making process for this next wave of computer-enriched technology approaches the saturation level of those of bar codes. Decision makers seem satisfied to use current metrics with the implementation process of RFID. However, the facts of the use of RFID may become the catafalque of their decisions. Since the 1960s, the evolution of computer-enhanced inventory tools, techniques, and decisions has increased the factors of complexity, creating nonmonotonically cascading risks, and changes to key performance indicators for human and machine systems. Communications as a logistics thread, in a disruptive environment or technology, has become an addictive change element: cell phones; handheld computers; and RFID data for decision making. This book focuses on these dependent variables in a constant discovery process for the best metric for decision makers using technology tracking elements and systems.

Traditional metrics

The metrics you choose depend on whether you are (or want to be) a global competitor in the international market or want to continue to be a domestic provider of goods and services. Size does matter when considering metrics for logistics and supply chain management.

There are several Internet sites to identify traditional performance metrics for logistics systems and supply chains. The metric for RFID is an information technology metric. And information technology can be categorized as having metrics that are functional or operational.[3]

The functional metrics are traditionally those of costs and derive from the accounting principles and financial policies and procedures of an organization, that is, basic accounting principles.[3] So, the cost of implementing an RFID system will be measured by the metrics that show up in the income statement, the balance sheet. The operational metrics are those that speak to the operations of the facility, the manufacturing plant, the revenue that is generated from making or moving something. It is all those performance metrics of capacity and capability.[3] In a materials management operation, the functional metrics might be the cost of the various processing functions, whereas the operational metrics would be those of the number of machines operating a lathe system or sorting system. The operational metrics are much easier to actually see, such as the height of the scrapwork pile, the rework rate, the number of parts per hour or per machine needed. Decision makers with or without the aid of RFID use both of these viewpoints to feel the pulse of the organization.[3]

As RFID moves into elements of the information infrastructure of the organization, many decision makers rely on the traditional metrics of cash flow, present value, future value, return on investment, and similar metrics. These metrics are holding up decisions today to invest today in RFID. But these metrics have played a dominant role in the past and will continue to do so in the future.[3] I do not want to belittle these measures. They are important, and as a business owner, I rely on them as key drivers of my business. However, there are other factors I consider such as positioning myself with my competition. Can RFID help get me more jobs or keep the cash flowing at a rate I want, but not any more for the investment? Does it really matter if I have RFID? The point here is that there are other intangible metrics, such as quality of the product or service, responsiveness to customer needs, and innovation to meet changing customer demands.[3]

When is the best time to look at a traditional metric for a decision for use of RFID in your business? You may not be able to find a suitable payoff right now with RFID because you are looking at that payoff, which returns at the wrong time. You may be looking, and probably are, as I often do, at the immediate metrics.[3] The actual benefit of an RFID system may not be with the internal metrics of function or of operations. These categories of metrics are really internal to the company. Going back to our systems view of the company, and considering the closed nature or open nature of the company, we also have metrics that are outside of the company.[3]

Metrics issues with time and temperature tags

RFID metrics for time and temperature seem quite simple. You want to know the temperature of a product as it moves along the supply chain within its chain of custody. One such RFID time and temperature tag is shown in Figure 4.2. This is an Alien Technology time and temperature tag that has been used to track the temperature of perishable products, such as fresh Alaskan wild salmon, from the waters off Alaska's coast to the fish markets in Chicago. There are many different varieties of time and temperature tags on the market. We used several in our laboratory.

In one experiment, using tags other than Alien, the original program for all tags was 6400 seconds, which means that tags would each wake up after every interval of 106.67 minutes and read the temperature. The tags have a limit of 64 intervals. So, the plan was to have students use these tags for a week of experimentation. This particular batch of RFID tags was new and had been certified by the vendor for accuracy and proper functioning. However, all these tags went off-line, using up their 64 intervals in less than 12 hours, prior to handing the tags out to the students for testing. That means the temperature recorded was in a time interval less than 106 minutes. Additionally, because all tags were together in a case being carried around by the instructor, we expected these tags to read about the same temperature even though the time intervals were shorter than expected. The results are shown in Figure 4.3.

Figure 4.2 Active RFID tag used to track time and temperature for fresh seafood shipments.

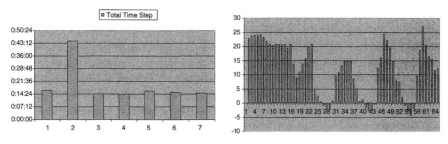

Figure 4.3 Time and temperature tag results that indicate a wide variation in time step shown on the left and an even wider variation in temperature recorded, while in a controlled temperature environment.

Temperature controlled chain of custody

The traditional metrics of a temperature controlled supply chain (TCSC) are cost and service, along with shelf life of the perishable product. Failing to maintain appropriate temperature control over products that move through the cool and cold supply chain can have severe costs in terms of brand image disruption or destruction, with the extreme being the death of consumers. But what exactly is TCSC? Smith defines the temperature controlled supply chain as "a food supply chain which requires food products to be maintained in a temperature controlled environment, rather than exposing them to whatever ambient temperatures prevail at the various stages of the supply chain."[4] The nature of the food products, such as meats, produce, or seafood; the legislative constraints; the temperature controlled technology used (or not used); and the transportation and distribution nodes along the supply chain help define the complexity of the TCSC.

Different perishable products require different temperature controls, and different regulations govern these controls in the global food market. The different temperature levels vary from frozen, cold chill, medium chill, and exotic chill.[4] The metrics of temperatures range from −25 to −18°C for ice cream and various food ingredients as a legal metric definition for frozen, to 0 to +1°C for cold chill, to +5°C for medium chill for fruits and vegetables, to a range from +10 to +15°C for exotic chill for eggs and bananas.[4] The management of these various temperature ranges for a variety of products is where RFID time and temperature tags can be most useful. The most complicated parts of the cool and cold supply chain, the TCSC, are the various nodes of transport, handling, and storage that are out of the control of the original producer of the goods and out of the control of the end consumer.

Smith reports that over the last 40 years, the United States and the U.K. have been experiencing a steady increase of at least 3% per year in the volume of temperature controlled products reaching supermarkets and consumers.[4] Items such as bagged salads are increasing, as ready meals and prepared salads are increasing in volume. In 2006, a news report cited evidence that U.S. iceberg lettuce and salad bags have been reported to contain possible contaminants.[6] The reason could be contaminated bags, the mishandling of the lettuce, or faulty temperature controls during any stage of production from harvesting in the field to placing on the supermarket shelves.

The use of RFID tags is already starting to have an impact on how the traditional supply chain works. Smith reported, "Technological changes in production and distribution have also allowed a transformation of the supply network."[5] The use of RFID tags in the international supply chains could bring better quality to the flow and use of foods and decrease illnesses due to mishandling of food and food products, but could also allow many independent distributors to enter the market. The logistics challenge in TCSC "is more formidable when the materials and products require temperature controls."[4] Speed, transportation, handling, accuracy, reliability, and storage take on more important roles. But this use of RFID technology comes at a cost, a very traditional metric. Time and temperature tags can run from $3 to $200 (U.S.). Being able to control where they are placed with the perishable items, inside or outside the product package, is still a question that seems to be answered on a case-by-case basis. Placing tags inside vans and refrigerated containers to check on the temperature flow patterns is also costly, but could be necessary to comply with local, state, federal, and international laws governing temperature controls for certain items.

In the past few years, "TCSCs are often seen as a specialist discipline within logistics."[5] This may be part of the drive behind a new metric to use with RFID in TCSC, that of the pace of the use of RFID technology. Looking at the growth of TCSC is a matter of statistics from different countries. In parallel, look at the growth of legislation and laws for food products. Now, enter the passive RFID tag, and since the Wal-Mart and Department of Defense mandate to start using passive RFID tags on pallets and containers, the pace seems to be increasing. There is still further research to confirm this metric; anecdotes are not enough to convince decision makers that their food handling process and carrier connections need improving. Paralleling this pace metric are other key metrics, such as packaging changes over the last few years, transportation container changes, shelf life of products, transportation time for goods, and cost of goods.

Three key issues in TCSC that could be affected by RFID and the related metrics for decision makers to pay attention to are cost of facilities and operations, food safety, and partnerships with RFID tracking and tracing of goods. Perhaps the most important of these three issues, from the standpoint of how RFID can assist, is food safety. When you examine the detailed food supply chain for use of RFID technology, you can identify the critical points along the supply chain that need reinforcing. Figure 3.2 shows the Alaska wild seafood supply chain, which already is sensitive to price fluctuations and unpredictable seafood harvests each year. By identifying locations along this supply chain, a structure can be designed to show how expansive the use of RFID tags can be in a cool and cold supply chain. Each point along this supply chain needs review for how the product is handled, the procedures, capabilities, assumptions, and vulnerabilities. These critical points are more important today because there is potential for sabotage along all points of any food supply chain. The Hazard Analysis Critical Control Points (HACCP) is a risk assessment tool that is useful for analysis at these critical nodes.

There are several critical nodes and events that need to be analyzed along the seafood supply chain in Figure 3.2.[4] First, examine how temperature settings are placed in a readable fashion on boxes, containers, load sheets, and run sheets. Second, examine how workers check and verify temperatures inside cooling or chilling or freezing containers. Third, examine how operations are handled between refrigerated storage containers and the loading dock before loading onto trucks or other containers. Fourth, ensure proper procedures are followed and documentation is updated as goods are moved through the distribution center or warehouse. Fifth, ensure that drivers of trucks understand the procedures and monitor temperatures inside the trailers, if a tractor-trailer unit is used. Finally, if all this manual documentation is handled by RFID technology, you may be able to skip most of these steps and rely on temperature readout devices that send an alarm to the loading dock, the supervisor, or the driver when the temperature bounds are breached. If the temperature goes below or above a present threshold, then you have added more value to the entire shipment and provided more security to the shipper and the customer.

When the use of RFID time and temperature tags is routine throughout all food supply chains, one possible benefit will be for the suppliers, the farmers, and the fishermen to concentrate on other aspects of their trade and look for other ways to add value to their products.[5]

Metrics illustration: warehouse record accuracy

Measures of performance or measures of effectiveness are two common metrical artifacts used by decision makers. In a warehouse, the usefulness of inventory records is more related to the accuracy of the physical count of items in stock. So, the metric for warehouses could be just that, a count, a number. This number will also have its second parameter of time, such as 10,000 items recorded this week or this month. Some lumberyards still conduct a monthly physical count of stock, which takes at least 1 if not 2 days, in any kind of weather. So, the metric we are looking at for this illustration is count per time. Also, as usual, there is another parameter of location. This, then, would be the count for a week at the Chesterfield, VA, warehouse for HT, a transportation freight forwarder, versus its Houston, Memphis, or Anchorage warehouse. Inventory records become essential to sound decision making, where there may be decisions made regarding the net requirements for a series of material items or release orders for material based on the accuracy of the available supply of materials in the local versus regional warehouse. The metric of inventory could be very precise and accurate, which is what the decision maker wants.

Look at the example of records in Figure 4.4. The aggregate inventory count is accurate when the record of the inventory is matched with the aggregate inventory of the items from both aisles. However, the accuracy from the two aisles is not reliable. The record count for part A01 is 100 items, but the count from the two aisles of parts is a combined metric of 102 items. As you scan the aisle counts, you could make an assumption or observation that the aisle 2 count could be inaccurate. The metric is still the count. But now you have a decision to make. Do you go back and see if the count on aisle 2 is as reported in Figure 4.4? If these numbers are not in hundreds but are also aggregate, and the true count is measured in the thousands, then the count of 100 is really 100,000. If this is inventory for manufacturing a product and the different inventory items are part of a multilevel bill of materials to make a product on a daily, weekly, or monthly basis, then the accuracy of aisle 2 becomes a major factor and possibly a bottleneck in process scheduling when parts go missing. Excess inventory is just another metric—the metrics of dollars per item of inventory to store and maintain in the warehouse. What to do? Because this is a book solely devoted to RFID metrics, one might jump at the assumption to tag every item on the shelf and have readers placed along the aisle ports or shelving spaces to give an accurate read of materials. However, the real problem might not be that easy to solve. The real problem may lie in just looking at the data.

Now, look at Figure 4.5. Here, we have an expanded inventory list of parts with four aisles of items needed to manufacture a product. The bill of materials may call for an equal number of parts for the assembly, and it is expected that all four aisles will always contain the same number of items. Looking at the data, you can spot a possible human error, or at least a suspicion. The current counting method for this warehouse may work just fine without RFID. However, if you examine the personnel conducting the inventory, you may find that Ashley has been with the company for 10 years, is very happy with her job, just got a promotion and raise, and loves her work. She sees a bright future ahead of her if she just keeps working hard

Part Number	Current Record	Aisle #1 Count	Aisle #2 Count
A01	100	50	52
A02	100	50	50
A03	100	50	49
S01	200	100	103
S02	200	100	98
S03	200	100	102
R01	400	200	200
R02	400	200	197
R03	400	200	196
R04	400	200	203
Total	2,500	1,250	1,250

Figure 4.4 Inventory record accuracy in a warehouse.

Part Number	Current Record	Aisle #1 Count	Aisle #2 Count	Aisle #3 Count	Aisle #4 Count
Inspector	Will	Ashley	Matt	Ashley	Matt
A01	200	50	52	50	53
A02	200	50	50	50	49
A03	200	50	49	50	49
S01	400	100	103	100	105
S02	400	100	98	100	100
S03	400	100	102	100	102
R01	800	200	200	200	204
R02	800	200	197	200	197
R03	800	200	196	200	188
R04	800	200	203	200	203
Total	5,000	1,250	1,250	1,250	1,250

Figure 4.5 Inventory record accuracy in a warehouse by inventory operator.

and producing accurate results reliably. The inventory clerk Matthew is a summer intern who thinks because he almost has a college degree that he should be in the front office being a manager apprentice or something like that; he should be making decisions, not moving or counting boxes. We may find the real problem is the person making the count. Or is that the problem? The result is a bad count on Matthew's task to conduct an inventory count. The real problem may be training; it may be a lax hiring policy; it may be hiring summer interns for a critical job without supervision. It may be several things. Could RFID help? The answer is a resounding yes; of course, you would not be reading this book if you did not think so. But is that the solution needed here? As we have said before, and will say many times, defining the problem is the critical point in deciding to use RFID, and a critical point in deciding what metric to use.

What metric is needed to resolve the problem arising from Figure 4.5? The metric count is important because of its direct impact on revenue and expenses, the cost of goods. But another important metric has to be the ability and capability of the human in the workforce. In this inventory example, we need qualitative metrics, skills, motivation, training, hiring process filters, and checks on task performance.

There are many types of reasons for the errors to occur in Figure 4.5. Poor records of inventory could be caused by unauthorized withdrawal of material; an unsecured warehouse and easy access to aisles 2 and 4; poorly trained people; an inaccurate process of creating the final inventory record; human errors in transcribing the data into the inventory database; lack of a feedback system to cycle count the inventory; and the amount of tolerance the company really allows in counting, such as 98% is determined acceptable.[2] The example in Figure 4.4 and Figure 4.5 shows a tolerance of between 2 and 11% of the physical count versus the inventory record. How much tolerance or variance can the company allow and still not interfere with customer needs, inside and outside the plant, and how does that tolerance translate to the metrics of expense?

Approach to developing metrics

Whatever metrics you use today or will use with RFID have to be acceptable to all levels of decision makers. Decision makers are looking for forecasts of profits, material needs, transportation problems, fuel increases, and many other factors. This is where operations research can become a key part of the metrics definition to meet the goals and objectives of the company.

Each logistics company, transport company, and supply chain integrator uses metrics. But not all companies that perform the same function use the same metric. Each freight forwarder does not necessarily use the same metric, due to the possibly hundreds of variables that could impact the cost of freight movement. There are three criteria for selecting the appropriate metric. We will get to creating new metrics later.

The first criterion is based on the system description of the business case, such as a supply chain integrator company. This is where the performance for that company is defined. The metric may measure various numerical or non-numerical parameters of the system. Of course, we will have already defined the system as closed or open. Besides numerical counts of inventory, number of trucks, and number of cargo planes landing each week, there are numerical factors that are ratios. A ratio is one of the strongest measures you can have, because it combines two parameters into one number. That number can be expressed as a fraction, as a percentage, or as a probability statement. For example, the percentage of 40,000 items of goods moved per year that were moved successfully, to the right place at the right time, could be 98.6% for the past year and overall 97.5% for the last 2 years.

In parallel, the probability of next year's metrics of measure of effectiveness to meet the needs of the customer could examine the last 5 years of percentages; using regression analysis, or other statistical analysis methods, you could predict that next year's measure of effectiveness will be 99.4%. Moving cargo by the ton or pound at 60 miles per hour by truck or at 600 miles per hour by cargo plane is but one part of a combined ratio of a metric for performance. Knowing where each item in the cargo bay or trailer using RFID is, is useful for a location parameter to add to this cargo movement metric. When we have metrics, for RFID, bar codes, or manual systems tracking, the choice of the measure as a ratio is essential for stronger meaning.

The second criterion is based on one of the basic principles we will explore in model building and process flow analysis. That is, when we use many different metrics and many different measures to describe the various components linked to form a supply chain management system, we must pay attention to that unit of the metric. There are many discussions in the academic community about how the metric units of the RFID world will not be any different from that of the bar code world. This book exists to differ with and to answer those academic pundits. When you are measuring ton miles per hour or ton miles per day, you cannot mix this metric with inventory per hour or inventory per day (if you were taking inventory that often). Also, you have to be very careful not to play with a ratio metric that is now dividing into another ratio metric. This is the classic one-number concept from my youth as a beginning operations research analyst in the 1960s and 1970s.

For example, you could have a ratio metric of 16,000 packages sorted per hour in a sorting plant in Anchorage. You could also have a metric of the cost of that conveyor system machine that is the backbone of the sort operation, with a measure of $100,000 per machine. Some may want to compress these two metrics into one number to get a stronger number. That is, you would divide $100,000 per machine by 16,000 packages per hour to get a new ratio of 6.25. Now, the metrical unit of this number 6.25 would be dollars per package per hour. You could interpret this as it cost $6.25 to move one package on this machine; a cost per box. Is this a valid measure of effectiveness of the sorting system? We can go further. Now we examine the 20 people who work this sorting operation in one 10-hour shift at a cost of $25 per hour per person. We can divide the $6.25 by 20 people and get 31¢ per person to move one package per machine cost. This sounds so logical, except you should really write down all the dimensions of each variable that is used to create each ratio and look at what it really means. Mixing ratios is meaningless, statistically. It may seem to mean something to some decision makers. I have fought against one-number concepts all my professional life of 40 years, and I will probably be doing so for the next 40 years.

The third criterion is simple: use the right metric for the right occasion. The occasion should be defined in terms of the company's objective of its supply chain business model. This metric must also be able to collect data that is considered by the decision maker and the statisticians as reliable and valid.

References

1. Drucker, Peter F., *Managing for the Future: The 1990s and Beyond*, Truman Talley Books/Dutton, New York. 1992.
2. Arnold, J. R., Chapman, Tony, and Chapman, Stephen N., *Introduction to Materials Management*, 5th ed., Pearson Prentice Hall, Upper Saddle River, NJ, 2004.
3. Devaraj, Sarv and Kohli, Rajiv, *The IT Payoff: Measuring The Business Value of Information Technology Investment*, Pearson Education/Prentice Hall, New York, 2002.
4. Hedgepeth, Oliver, classroom assignment, UAA, January 2005.
5. Pamphlet No. 71.1, Force Developments: The Measurement of Effectiveness, U.S. Army Combat Developments Command, Fort Belvoir, Virginia, 1973.

chapter five

Developing a few new metrics for RFID applications

"...but, there was much to grok."[1]

Robert A. Heinlein

Supply chains, metrics, and transportation

The future of supply chain transportation is being held hostage by the uncertain economic model and mandated applications of RFID technology. Not since the 1960s revolution of computer technology swept away office workers and rearranged the manager's entire concept of data management requirements has there been an impact on manufacturing, distribution, and retail and wholesale business as will happen with the current wave of massive, ubiquitous, passive RFID tags and different levels of aggregate active tags, and related software integration.

Decision makers who use a monthly physical count of inventory may continue with this decision, or they may watch their inventory rise and fall on their computer screen, second by second, if they use RFID. The amount of potential data increase may become significant, maybe overwhelming amounts. However, the supply and demand for this new streaming inventory data based on RFID could have an impact on transportation, manufacturing, and distribution costs, and the flow of information within the retail and wholesale supply chain system.

How do we measure this impact before these massive floods of data from these unending applications of RFID? Projections of the computer's impacts from the 1960s into the next decades and beyond fell apart due to not anticipating social and cultural impacts that technology brought to management. What we examine today with RFID is not a trend analysis for the applications of RFID; it is a step-by-step method to identify those quantitative metrics that will be impacted most significantly by the business applications presented earlier using RFID technology. The format for these quantitative metrics is presented so you can calculate and refine these measures to fit your specific transportation strategic needs when, not if, RFID technology invades your domain.

To measure a bar code you need to count; to measure an RFID code, you need to think. And the simplest thinking is that you have a capability with RFID tags to have better inventory management and transportation management of those goods. But is this too simple? Another question many trade articles entice the reader with is, What is the real cost of using such RFID technology, especially on the transport side of the supply chain and logistics management?

Transport supply and demands

Transportation and communications are not mutually exclusive. The use of communications can replace transportation. With the use of the Internet, orders for goods are placed, trucks and trailers and drivers are contacted, and goods are traced across the landscape to the final destination, just to make sure the load arrives on time at the right warehouse. Time is money. Transportation is money. Communications is money. All three are linked and more so if we have smart, RFID-laced cargo that can be read along the transportation route so the customer, the carrier, and the shipper can see the shipment move from point to point. So, one economic metric for RFID is the economic model of trading time for money, all based on communications of track and trace of goods, and the physical process of movement of goods. If we have better visibility of the load in movement by using RFID tags, and the cargo is recorded the second the tractor-trailer enters a warehouse yard, and the goods are instantly recorded the second the forklift operator drives the pallet out of the trailer and places it on the warehouse floor, then we have a model for many RFID metrics, which are all traditional, to be more efficient, and the process to be more effective. If this all works out, then the trailer can be seen as empty as soon as physically possible and can be made ready to pick up a return load faster. Time is saved; money is saved.

Since 9/11, the increase of over 600 new rules, laws, and other barriers to crossing into U.S. ports has impacted transportation, slowing it down. Figure 5.1 shows an active RFID tag device placed onto an ammunition container; the military has been using such active tags not only to identify the contents, but also for tamper-resistant uses. There are procedures for known shippers to clear their tractor-trailers past Customs and border checks and continue on to their final destination. But as is seen on the news and in every newspaper, almost weekly, there are over 17,000 containers entering the United States each day, and it seems a small percentage, maybe as low as 4%, are actually being inspected. And with the possibility of the post-9/11 entry rules not going away, the search for technology solutions, such as RFID, to keep the transports moving continues and is expected to continue

Figure 5.1 Photo of an ammunition cargo container with an active RFID tag placed on the outside.

for the next few years. So, what wireless partially did to the land phone, RFID is doing to transportation visibility.

The highways, skyways, railways, and sea lanes of the United States are the blood vessels of the body business logistics. The blood that flows along these vessels and arteries are the trucks, the trailers, the rail cars, the ships, the barges, and the air cargo planes. Transportation is essential to all that business stands to gain; without transportation, the purpose of the supply chain systems ceases to have meaning. The RFID technology is fast becoming a part of that flow, a new life force within the transportation system, and transportation metrics.

Globalization before 9/11 seemed to be on a steady growth of connecting larger supply chains. Post-9/11 has placed constraints on global supply chain systems. Supply chains that were easily defined as open now seem like closed systems with new boundaries. However, despite the 9/11 impact on supply chains and transportation, manufacturing seems to be flourishing in countries like China. Daily newspaper articles on the trade deficit between the United States and China tell the story of significant flows of material between these two countries. My visits to the port cities of China in 2005 and 2006 attest to the almost uncontrolled expansion of ports, and China's move into the RFID market indicates that the Chinese see the RFID metric as part of their transportation and manufacturing future. Transportation seems to be an almost visible and invisible force that is allowing goods to be purchased around the world, especially in the United States, for cheaper prices than if they were made in the United States. The success or failure may rest on getting the goods to market just in time; will RFID create a new metric that is a little faster than JIT? The RFID tag does seem to now be part of both the visible and invisible market forces. With billions and billions of RFID tags just around the corner on not only pallets of goods, but individual items, like your Gillette razor, RFID technology is creating a tidal wave force in transportation economy.

In the early part of this century, transportation seemed simpler. It was a network of roads and lanes, crossing each other, interconnected, with multimodal transportation a simple process. Logistics or transportation managers would map out routes for transporting goods from one manufacturer toward the end customer. RFID technology is part of the new communications arm of transportation.

There are several growing trends in the importance of transportation in this new era of microsensors, all of which will be affected by RFID technology. These trends are increased demand for transport for freight and passengers; a reduction in the cost per unit of goods transported; and the expansion of all kinds of transportation infrastructure. With the number of people and goods being transported comes the need for more accurate data to get the right goods and people to the right place at the right time. RFID tags are being used for many transportation operational functions, such as for luggage at some U.S. airports and smart tags on the windshield of your car as you drive through the smart lane tollbooth on the way to work each day. This could be a growing trend that more goods will be transported into and out of increasingly geographic-challenged areas. And the customer is demanding goods and services, no matter whether they are destined for a cabin in Alaska or a condo in Washington, D.C.

The metrics trap coming with RFID

RFID technology is the new yardstick for transportation costs. Transportation is complex, following the principles of a nonrandom network, which simply means the patterns of supply chain and transportation are not random events. The smart systems based on RFID technology will and are starting to display self-organizing behaviors, which is something Henry Ford could not understand, but which systems thinkers do understand. This nonrandom nature of the RFID supply chain system contributes to the inability to accurately predict the transportation demand and measure its impact on more than transportation. So, the RFID metric for the cost of transportation may include information on where goods are located along a supply chain network, the price of those goods as transportation costs fluctuate, and contract boundaries on the enforcement of laws governing the movement of goods. There are at least three RFID metrics for transportation. They can be categorized as follows:

- **RFID Metric 1: Location and Distance.** Location of the goods and the transportation carrier is basic to calculating the various distances the carrier moves the load; the variable costs, such as fuel surcharges and tolls; and weather constraints. This location is given in points along the route by the RFID indicators, the distance along the segments of the route, and the total distance traveled with a load and return, with or without cargo.

- **RFID Metric 2: Multiple Modes.** With RFID applied to the cargo and the trailer or container, the different modes of transport can be accurately recorded, along with time and temperature data for updated insurance underwriting. The cost of each mode will be different and the cost of insurance will vary based on history or the use of RFID technology to offset past history.
- **RFID Metric 3: Purchase and Freight Forwarding Costs.** With RFID, the cost of purchasing goods from different vendors could change, with negotiating rules and policies linked to vendors that support RFID along with cargo or containers, and with use of RFID at cross-docking facilities.

These RFID metrics are based on at least two concepts for transportation systems:

- **Transportation network nodes.** The supply chain system of today is a complex network of highways, distribution centers, ports, ships, airports, freight handling companies, transport companies, and supply chain integration service providers. Network analysts, operations research analysts, and industrial engineers are all familiar with how to define the optimum transportation network and nodes needed to support a supply chain and logistics operations, for industry and government.
- **Transportation network demand.** The force and the engine behind the force of the transportation network are the various demands: There are customer demands, carrier demands, supplier demands, manufacturer demands, materials management demands, and boundary constraints (customs and border rules) on all these demands. The ebb and flow of the force along the transportation network varies from node to node. There are different varieties of the dimensions, the costs factors, which control this variety.

These two concepts provide a variety of systems perspectives for transportation analysis to control the flow of goods along the various logistics threads of the network. There are many operational research methods to analyze these threads along the transportation network. A simple method often used is graph theory, which examines the supply chain network as a series of arcs and nodes, the logistics threads, needed to be taken by the containers carrying the goods to be tagged by RFID for tracking and tracing. There is over-the-counter software to aid the analyst who is not a trained operations research analyst. This has been a simple mapping exercise over the last 20 years, thanks to the computer, and since the Internet, it has become much easier to analyze networks. A transportation route, tolls, bridge heights, roads that cannot take trucks, and much more information is available today. The RFID data entry into this internet of things today will start to give better understanding to the once-twisted logistics threads that resembled a football stadium full of spaghetti. The variety of the nodes and threads is complex and could look chaotic. However, RFID could help reduce much of the uncertainty of optimizing movement along these network connections. In order to better understand the complexity of this process, you have to also consider at least the following key variables of transportation:

- Network system components
- Network logistics threads
- Network variety
- Closed networks
- Open networks

There are many dimensions when determining the cost of transportation. There are many known, seasonal, political, and unknown dimensions. How you decide the common denominator, as a metaphor, of any metric is dependent on understanding today's changing supply chain, due to RFID. In the transportation world, there are already various kinds of rates. Some are fixed, such as public transportation. But for freight, these rates are negotiable. We are not going into the detailed transportation requirements in this book. You just need to know or be aware that cost will probably vary depending on the scale of the movement. Given political considerations, fuel costs, and impacts of significance such as the Iraq War, transportation costs could vary widely.

Reference

1. Heinlein, Robert A., *Stranger in a Strange Land*, Doubleday, New York, 1961.

chapter six

War game
planning for RFID

"...the greater the divorce of producer from consumer ... the more the market ... with all its hidden assumptions ... came to dominate social reality."[1]

Alvin Toffler

Assumption-based planning

Assumption-based planning (ABP) was developed by the Rand Corporation in 1989 to deal with the uncertainty in the U.S. Army policy-making and planning part of going to war with Iraq.[2] Since then, oil companies have used it to understand the uncertainties in oil exploration. Assumption-based planning is a measurement tool that can be used for any large-scale project such as implementing RFID into the Department of Defense's (DoD's) logistics and supply chain systems, a system of old and new technologies that do not communicate well with each other.

ABP is based on systems thinking and forces you to think about all the parts or components of your system, such as material management, inventory control, transportation, marketing, and customer service. ABP is a management tool as well as an analytical tool; because it is not mathematical, it lends itself to easy use. But the user must understand when to use this tool, or it will be no good. It is like a yardstick; it is great for measuring distance, not so great for measuring the size of a cloud. You will be faced with understanding the factors of your logistics system, the changes being brought to its internal and external processes by RFID or other automatic identification systems, and most importantly, the interactions of the data produced by using RFID. ABP is a tool that is used to measure first, and then compare with what is expected, and then prepare a response to what you find. Using ABP will be different from any other technology forecasting or trend method you have ever used.

There are five basic steps in using ABP.[2] First, identify only the important assumptions that govern or control your system, your logistics system, your supply chain, your warehouse, and your purchasing policy. Second, identify the vulnerabilities to these important assumptions. That is, all assumptions will perish over time. The trick is to be smart enough to find out beforehand when your assumption about your business will fail, because your business will possibly fail right along with the assumption. Third, identify any signposts along the business path that you are taking. With RFID, the Wal-Mart and Department of Defense's decision makers have made some basic assumptions regarding the implementation of RFID for its major suppliers. There are vulnerabilities to those assumptions. But just as important, there are road signs that you can spot that may impact or invalidate the vulnerability or assumption.

This is where ABP starts to get really interesting. A signpost could be something as simple as the war with Iraq. That signpost could lead to the assumption that there is a need for total visibility of all military supplies entering the war zone. Or there could be the signpost of a hurricane in Florida and what it means to the assumptions for the cruise industry or for Home Depot. Fourth, there is the shaping action, which is some action plan or plan of attack to make sure the signpost does not invalidate the vulnerability or the assumptions. You must take some action when you see a threat, hurricane, or war. Fifth, there is what is called a hedging action or what we like to call *Plan B*. That is where you are going to lose some major customer, and you had better figure out what to do to keep your factory open.[2]

We first need to define *assumption*. It is an assertion describing a specific characteristic of a future state of events, which is the foundation of your organization's process, policy, and procedures to conduct business and make a profit. For example, an assumption might be that the academic year at a college is divided into two semesters. Or that the college's Russian student population will grow by 10%. These assertions are not just wishful thinking. They are statements based on observable facts. If you go to the college catalog, you will see that the year is divided into two semesters, fall and spring. If you go to the registrar and look at the Russian student population, you could see that over the last 5 years, the trend is about 10% growth of new Russian students; then, such an assumption is probably true.

The importance of the assumption is essential to the ABP method of measuring the economic health of your company. You can measure the importance factor if you examine this assumption and consider what would happen to your company if this assumption were not true. Now, this is all subjective. And judging the importance may take a group of people to evaluate, just like the validity of the assumption.

The vulnerability of the assumption depends a great deal on the timeframe for the assumptions. When your company began, it had some basic assumptions. Are those assumptions still working today? When the assumption became vulnerable has as much to do with how long the assumption has been stable as it does with what external or internal events are attacking that assumption.

To begin to identify your assumptions may take a team of people. You can conduct brainstorming sessions, goals analysis, causal analysis, game playing, or Delphi methods. Any method will do, as long as the important assumptions are identified.

ABP is an analytic approach that is most useful in a structured business setting, such as the Department of Defense or Wal-Mart distribution system. When the system you are describing is messy, as all large systems tend to be, ABP will provide a structured framework for planning your way around the elements that are causing the messiness to occur. It is extremely useful in uncertain times, when uncertain results may occur from a massive adoption of some new technology into an ongoing business process, like RFID.

Business transition war game for RFID

This war game has been designed specifically for the RFID needs or demands in the logistics and supply chain management community to be used in upper division undergraduate and graduate courses, as well as business and political decision makers. This business war game is designed as a major exercise for students as part of the terminal course of instruction. As such, participants are expected either to use their knowledge gained from the previous courses of instruction, or to be middle to upper management in business or government.

This war game has been specifically designed for understanding RFID as a tool in the transformation of state, national, or international business enterprises within the constraints of the respective governmental and organizational infrastructure and environment, as well as the RFID-related issues and problems facing its growing relationship in the global economy. The results of this game should form the basis of dialogue for supply chain and logistics processes, the dynamics of supply chains and logistics operations, and emerging complexities facing the governing bodies within individual and free-market enterprises that are growing at a fast pace. The ability of these new markets to better understand how to compete within the local and the global regions will strengthen the market penetration potential of businesses in your state or business sphere of interest. Therefore, this business war game is a decision-making tool for strategic planners using RFID technology or planning to use RFID technology. This war game has been applied in Alaska and in Russia as part of an executive master's degree in global logistics management. Results for its application are presented.

What is a war game?

This chapter is a war game; it is not the development of a business plan for RFID. A business plan lays out the detailed steps to establish a funded, viable business enterprise built around RFID technology. As a war game, the chapter will focus on the opportunities and the obstacles that lay on the path toward building RFID into or as a new business venture. Only after the war game is completed can a business plan for implementation of RFID be developed to address the steps in implementing RFID as part of a new business technological infrastructure.

A war game is a contest between two individuals or groups of decision makers, each of whom wants to win. The purpose in this war game is to examine the optimal strategies or series of potential strategies formed in the conflict of creating RFID innovation in business, and creating and sustaining a business in an environment that is full of conflict in the use of emerging technology. The war game members are part of a group seeking to develop a business case for RFID. The group wants its business to "win" and survive for many years, make a profit, have customer loyalty, and provide quality products and services.

So, who are the two sides for this war game? If you are considering running an RFID pilot test or have decided to implement RFID as part of your information technology solution, then you and your organization are one side. The side that is working to undermine your organization is made up of those factors within the supply chain, within the logistics systems, within the culture of your business community, and in the environment that surrounds your business system, all of which seek to stop you from being successful. You and your organization have a face, a brand, a name, and are clearly identifiable. The forces you are working to beat in the war game are many faceless individuals and groups and systems. So, this is almost like a war game where you are fighting a terrorist organization, or the aftermath of an earthquake or tsunami (as happened in Asia in December 2004), or the continuing problems with business and human suffering from Hurricane Katrina; instead, it is the terrorist or tidal forces of anti-new-business-ways, sometimes simply called change, but today we call it the RFID mandate.

You and your organization will face many threats to your business decision to implement or even test RFID. Your goal is to minimize your maximum losses by making decisions that have supportable assumptions, that have objectively observed or well-thought-out (that is second-order cybernetics) vulnerabilities, and plausible actions to defeat these vulnerabilities.

War game for Alaska and RFID

At the time of publishing this book, this war game has been exercised for 5 years in Alaska. In considering the use of RFID for Alaska, the decision makers who play this war game have to understand the market forces within Alaska, which currently are constrained by limited lines of rail transportation linking goods and services within and outside Alaska and lack of roadway infrastructures. This war game allowed practitioners to explore the metrics from mental models, performance measures, and basic assumptions surrounding the development of a global logistics business initiative. The goal was to provide insights into those factors that define the goals of a new logistics business venture in the Alaska (or your state's) business community and open the dialogue to explore options to address bottlenecks in pursuing a successful international business.

War game learning objectives for Alaskan RFID business case

The objectives of the war game were, and are, repeated many times: to explore an RFID technology-based opportunity within Alaska. Specific objectives are listed below:

- To transform Alaska (or your state) into a successful global logistics partner
- To analyze the assumptions for this transformation by identifying which assumptions could be faulty; and make recommendations for new assumptions
- To explore global business communications and control processes of the workforce and the technology needed for an Alaskan logistics or supply chain business to compete successfully in the global marketplace
- To identify bottlenecks, issues, and problems that could limit this transformation process
- To identify potential solutions to the bottlenecks, issues, and problems
- To develop global business strategies, requirements, and resource needs for the development of this logistics business

War game design

The war game is usually designed in two parts. The first part consists of senior stakeholders in Alaska, much like an economic development organization. These stakeholders are concerned with the strategic, policy, and top-level process implications of developing global logistics enterprises within Alaska based on use of emerging RFID logistics and supply chain technology. The senior stakeholders session usually explores the issues and problems prevalent in Alaska in the current timeframe and that may exist for the next 5 years. The senior stakeholders are to discuss and identify the issues surrounding the following goals as a final product:

- To describe the emerging local and global economy that is facing Alaska today
- To develop a set of useful metrics or performance measures that can be used to track the progress of Alaska in this new global economy
- To use these performance measures to identify and describe the status of how Alaska measures up to these metrics today
- To provide a set of recommendations for the Alaskan economy that will become the new strategic plan of action for Alaska, listing specific recommendations needed to have a successful action plan

A second entity of the war game consists of business developers, called Senior Business Leaders. These individuals are concerned with the tactical implications of actual implementation of an RFID business venture within Alaska. Thus, you may have surmised that each cohort is divided into two groups: Senior Stakeholders and Senior Business Leaders. To this end, the participants will be expected to address their business case from the following questions:

- What are the key RFID challenges facing logistics today?
- What are some difficulties you would expect to encounter in trying to measure supply chain performance with RFID?
- What are the possible bottlenecks that may be causing a problem, and possible methods for solution?
- What are the greatest areas of opportunity of using RFID for logistics operations, and why did you choose this area?
- How do RFID improvements in logistics productivity affect the economy as a whole, as well as the position of individual consumers?
- What are some of the difficulties you would expect to encounter in trying to measure supply chain performance with RFID?
- What are the major obstacles to successfully implementing an RFID-based supply chain management system?

War game dynamics

The game is designed to go through a series of steps over a period of several months (it has been performed in a matter of weeks, but the dynamics of the group interactions have to be curtailed). Each step in the process is to be conducted in a series of public debates or discussion groups, and a series of smaller group discussions, to address the war game objectives.

War game schedule

The following detailed schedule of tasks is presented with approximate times that have been used in previous, similar business war games. The times are offered as a guide to gauge your activity; however, we have found that some groups have spent many more hours or days on some steps. The total amount of time depends on the thoroughness of your thinking and digging for the best answers to some tough questions.

Phase I

- Step 1 (1 hour)
  Introduction of the war game by instructor
  Assignment of participants into different cells
  Review of current and new Alaska (or other state or country) economy
  Distribution of Players' Guides
  Selection of a War Game Director by each business cell
- Step 2: Begin Phase I
  Both cells review Players' Guides and begin play in parallel (1–2 hours)
  Instructor works separately with both cells for specific questions

Task 1: Review war game learning objectives. Determine if these objectives are complete. If not, then discuss with the instructor (1–2 hours)

Task 2: Answer five survey questions about Alaska (or other state or country) (3–5 hours)

Task 3: Develop a mental model map linking from the answers given in Task 2 (4–5 hours)

Task 4: Develop assumption-based planning structure (6–10 hours)
(Note: This Step 2 has taken several weeks with one cohort.)

- Step 3: Separate presentations to instructors by each business cell (1–2 hours each, separate times)
- Step 4: Instructions for Phase II (1 hour)

Phase II

- Step 1: Review of Phase I results. Presentations by both business cells (3 hours)
- Step 2: Begin Phase II
Task 5: Revise mental models (2–3 hours)
Task 6: Develop measures of performance (3–4 hours)
Task 7: Define Alaska's (or some other state or country's) hierarchy of needs (4–6 hours)
- Step 3: Develop final presentations and written reports by both groups (2–3 weeks)
- Step 4: Presentations by both groups (2-hour presentation each)
- Step 5: Post-mortem of war game (1–2 hours)

Task to be performed

As mentioned in the above schedule, there are five basic questions that need to be answered in some level of detail. Each person in the group provides an answer. When everyone has completed his or her answer, they combine and share answers. Then, the group compiles one set of answers for each question on which the group can agree. The five survey questions used for the Alaska war game were

1. Within your (this is your own organization, not the war game) occupational area, what is your view on Alaska's logistics management skills today? Why?
2. How do you think that the Alaskan profession of logistics management will change in the next 5 years? Why?
3. Within your organization, what do you believe is the greatest difficulty facing today's Alaskan logistics manager? Why?
4. What do you think is the most critical training area that logistics managers need to perform their job better? Why?
5. What are some logistics management tools or processes that are very successful for supply chains within Alaska?

Developing an assumption-based planning structure is the central focus of this war game. The five steps of ABP are followed, one step at a time. The first step is to identify important assumptions. These assumptions are those that will be made to determine how the businesses perform or would like to perform in Alaska for the next 5 years. Remember that an assumption is a statement, not a goal, that says something about the future that could impact business and the business of logistics and supply chain management in this war game. An example of an assumption of this kind is, "The United States will continue for the next 5 years to implement its new smart border rules, many of which are RFID-based, for importing goods." Notice that there is a timeframe on this assumption. Remember, all assumptions will become false at some time in the future. Your job is easier in that the timeframe from all your assumptions is the next 5 years. At this point, the cohort is asked to determine five load-bearing or important assumptions; as many as 40 assumptions were created by one cohort, but quickly pared down when they realized the enormity of the war game.

Step 2 is to identify the assumptions' vulnerabilities, closely linked to Step 3, which is to identify signposts affecting these vulnerabilities. These signposts are those events that, when large enough, clearly indicate a change to these vulnerabilities. The future is nothing but uncertainty. However, if we are prepared for possible alternative future events, then we will know how to react to that uncertainty when it does arise, instead of being unprepared and surprised. For example, the rising use of cell phones was not predicted, but was expected by some people. Those people are rich now. This is not an easy task; there may be only a single signpost that affects all assumptions. Such was the case when there were signals in 1985 that the Soviet Union was about to break up. Such a possible event would affect everything from the assumptions of the Cold War to adding the Russian Far East to the global logistics marketplace, where it did not exist before that event took place. A recent signpost is the Asian tsunami disaster in December 2004. How did this event affect the global and Alaskan seafood supply chain?

Step 4 is where the groups define shaping actions. You use shaping actions every day of your life. One saying or shaping action is, "An apple a day keeps the doctor away." A shaping action can be positive or negative to your basic assumptions and vulnerabilities. A shaping action can prevent your assumption from being true for the next 5 years. Or a shaping action may be one you see that will reinforce your assumption and really cause it or be used to shape it into becoming a reality. This shaping action is about something that has not yet happened; it is a possible future event, again in the next 5 years. To help you identify these shaping actions, answer these three questions:

- How soon may the assumption fail (within the 5-year timeframe)?
- How well can the violation be seen (within the 5-year timeframe)?

- How much time will be needed for some kind of action to take place if a violation is to occur (within the 5-year timeframe)?

The outputs of the cohort's discussions while answering these questions will provide a sense of any shaping actions that may be needed.

Step 5 is to define hedging actions. Previously, the signposts gave you a tool to measure and monitor the world of Alaska. The shaping actions gave you a measuring tool and sense of how to change this Alaskan world in the next 5 years to be positive to logistics and supply chain management businesses. But what if something totally unexpected happens that is out of your view, that is out of the blue, that is so totally unexpected you cannot think of it now? This is addressed by the hedging action. These are those actions that would make RFID in the logistics and supply chain management business in Alaska work despite the world being out of control. This hedging action is some action you could take to prepare the organization of such logistics RFID activities in Alaska for the upcoming failure of one of the important assumptions.

You can discover your hedging actions by starting with the vulnerabilities you defined earlier. These hedging actions are not opinions like you used for all the other tasks. They are actions to take regardless of the desirability of the world events affecting Alaska. A simple example is making plans to move your picnic indoors just in case it does rain this afternoon, even if you are not looking forward to the picnic at all; you may not want to go to the picnic, but are being forced to by your boss, husband, or wife. You do not really care if it rains or not. Either way, you still have to go. So, the hedging action is something to be done before the rain comes, such as clearing out a space, getting chairs, moving the charcoal grill inside. None of these actions will keep the rain from coming. So, defining the hedging action requires you to see the situation where the assumption is violated, for example, maybe a rain shower in the picnic area.

Developing RFID metrics or measures of performance occurs after the ABP analysis is complete, the first time. You actually have to go back through the process twice. This is a forced feedback approach that ensures that the cohorts understand feedback and, after examining a mental model from Phase I, do eventually see the value of revisiting the process in Phase II. The cohort then develops a set of useful RFID metrics or performance measures that can be used to track the progress of Alaska in this new global economy based on RFID or smart container technology.

The cohort then uses these performance measures to identify and describe the status of how Alaska measures up to these RFID metrics today as decision-making planning tools. These RFID metrics will then be used to define a hierarchy of needs, one for the state of Alaska and one for the senior stakeholders' business venture into the world of RFID. This is a list of the strategic needs of Alaska, your state, or an organization over the next 5 years. This hierarchy of needs is a list of strategic events or policies or programs that you think should be implemented or continued or changed, to ensure that in the next 5 years, Alaska's economy continues to grow into the global logistics, global supply chain, and global economy as a productive and prosperous partner.

Assumptions for RFID

Bar codes take a manual process and automate it, reducing human error and increasing the speed of data entry, with or without a human operator. Bar codes speak five or six times in a product's lifetime. RFID codes can speak 200–1000 times a second for that same product, or as often as each supply chain event happens. The end user searches for a bar code, but the RFID code searches for the user. These are some of the metaphors we read about almost daily concerning the rapid advance of low-cost RFID tags. The success of RFID seems a given, but the true price is yet to be determined.

Wal-Mart and the other leaders of the current low-cost RFID movement see great benefit for the price in their use to improve efficiencies in logistics management and supply chain flow of goods and information and management. But the burden of using RFID is on the suppliers to budget and invest in this technology, not the Wal-Marts of the world. The hundreds of articles on the potential benefits to the suppliers are, however, just assumptions or beliefs. But these beliefs are based on serious thought and analysis. However, assumptions, beliefs, and promises can still cause trouble if sufficient "what if" analysis is not performed on these assumptions and beliefs.[2] One such promise that we are counting on from greater visibility will be in the reduction of overall supply chain waste. Many of us have played supply chain games to depict what delay this visibility of the supply chain can do to inventory levels. Demand forecasting fluctuates wildly and goes out of control. The ability to capture an increase in accurate and timely shipment data should be a significant benefit for the database administrator and for demand forecasting.

It is assumed that RFID would allow shipment data to be more easily corrected for errors. Although RFID benefits are easy to find, the full impact of RFID is likely to be 5 to 10 years away, because the RFID programs in place today are merely pilots or tests. With such a timescale of nearly a decade, and given the advent of technology improvements over the last 10 years, many unknown benefits and problems will arise. We are still early in the business of using RFID for end-to-end business. There are many benefits, but also many unknown unknowns—we will refer to this as a pitfall. Even where RFID is being used, not everyone understands its impact. Interestingly, a recent personal interview (in September 2004) with employees of one retail store using RFID on cases and pallets showed that the employees had no idea what RFID was and how it was being used to improve their backroom operations. Analysis of benefits and costs of RFID has to take into consideration many assumptions, beliefs, and promises. Scholars are working on these pitfalls now, with theses and dissertations growing in number. In the meantime, the business of using RFID technology also grows.

RFID tags are becoming a process change element on products within the logistics systems reaching from Alaska seafood processors, to tire manufacturers in North Carolina, to thousands of giant retail stores like Wal-Mart. We want to explore where these pitfalls of RFID are located in the logistics landscape, who or what is creating them, and whether or not we can avoid them. Can we step around the pitfall, or is it too big to go around? Or can we detour along the path?

In looking for a pitfall, technology history and my experience tell us there are unknown unknowns, and this is the case in the use of RFID. Given the rich history of introducing technology into society over the last few generations and especially the last 50 years—the computer generation—we have the following assumptions:

Assumption #1: There are pitfalls, or unknown unknowns, in the use of RFID.

Now, how do we find these unknown unknowns or pitfalls? The search for where such a pitfall may be coming from is to search for a suspect: who, or what, may be creating the pitfall.

One such suspect could be the antitechnology force at work in our society, which has been with business since the early days of the Industrial Revolution.[3] There are decades of studies on the causes for, and identity of, those people or groups of people in the antitechnology movement.[3] The technology could be as simple (to us) as the use of electricity, the invention of the telephone, the invention of the typewriter, or the automobile. These antitechnology forces can be a group of people who see their way of life threatened, their power diminished, or their jobs lost. Industrial robots working on the factory assembly line in the last 50 years were just as much a crisis for some factory workers as the early automation that brought in the Industrial Revolution. The forces against technology may be more than the few combatants picketing, loudly protesting against the technology insertion into their job. These are the stakeholders of whom business must always be mindful. So, we have the second assumption:

Assumption #2: There are antitechnology forces against the use of RFID.

The success of bar codes linked with this growing storage and processing speed and growing communications capability of the Internet and computer products is still unfolding. Part of that story is now the RFID dimension. With the martial law-like mandate from Wal-Mart and the Department of Defense, the changes will continue, but at what pace, at what price, and for whom? One might easily suspect Wal-Mart or the DoD as some kind of force causing the antitechnology voices to be raised. But are they this force, or are they merely acting on a natural progression of technology, of the next new novel concept that was going to happen anyway? And the third assumption is:

Assumption #3: RFID is part of a natural progression in technology.

It seems that a near-crisis feeling emerges before some new concept, like the current passive RFID tag mandates. Martino helped us understand what the impact of the last 50 years of computer technology had on business and our daily lives.[3] He described how technological adoption changes our expected values or models or assumptions about our life. If you have read the hundreds of trade magazine and newspaper articles on RFID, you may have begun to glimpse the paradigm shift that is likely to occur in the retail industry, first the manufacturing plant, warehouse, and trucking company. Wal-Mart and DoD expect a paradigm shift.

When we examine what Martino said about the past technology evolution and acceptance of the paradigm of the bar code, we should try to find if there is a crisis in the use of the bar code. Yes, the newspapers are full of articles of how many billions of dollars will be saved in warehouses, manufacturing, material management, transportation, retail stores, product shrinkage, and security by using such RFID tags as a replacement for or in addition to bar codes. But the business of business and the business of engineering management and systems engineering have been streamlining and minimizing cost factors along these logistics and supply chain systems at an increasing pace during the last 2 decades. So where is any crisis? The assumption is:

Assumption #4: There is no crisis in the use of bar codes.

RFID will not be a new paradigm that happens overnight. Over the last 40 years, my experience indicates that the newest computer technology of RFID will overlap the old technology of bar codes for some time. It is like a process of collecting and storing the same data, but in a different way. And RFID tags are doing just that; they are handling the same bundle of data that the bar codes handle, at least for now. Currently, RFID is destined to coexist with, not replace, bar codes on containers. The bar codes' demise could be suspected of becoming another urban legend, like the paperless society. The assumption is:

Assumption #5: RFID will not replace bar codes.

So we have a snapshot that the use of RFID will not replace the simple bar code. Because there is no crisis in using the bar code, the RFID has many unknown unknowns in its use. It is part of the latest natural progression of technology that is going to happen, even if there are those people who will fight against it. There seem to be suspects everywhere.

But these are not all the assumptions. What are all of the unknown unknowns or pitfalls in using RFID? There are several steps you can take to better understand the full range of the critical assumptions at work with the RFID system. First, gather your team and identify and then challenge the assumptions for why you are using bar codes and why you should use RFID tags. Second, look for possible reasons why and when the assumption can be wrong, because assumptions within an organization tend to be vulnerable to outside events. Third, whatever you found that made your assumption vulnerable, recheck often that vulnerability as well.[2] Finally, dredge up those lessons learned when you first started implementing bar codes. You may find that the same lessons learned there apply again for RFID.

Did we find out who the suspects are in the unfolding story of RFID? At first, they appear to be those who are leading the mandate: Wal-Mart and

DoD. But to be fair to these organizations, the key suspects are the assumptions. By challenging these assumptions, you can avoid a pitfall or find a detour around one. By challenging these assumptions, you will find the suspect in your own organization's RFID implementation plan. The pitfalls or the unknown unknowns are real; they are often overlooked, but they can be found.

References

1. Toffler, Alvin, *The Third Wave*, Bantam Books, New York, 1980, 202–211.
2. Dewar, James A., Builder, Carl H., Hix, William H., and Levin, Morlie H., *Assumption-Based Planning: A Planning Tool for Very Uncertain Times*, Rand, 2004, 5–46.
3. Martino, Joseph P., *Technological Forecasting for Decision Making*, McGraw-Hill, Inc., New York, 1993, 93–114.

chapter seven

RFID forecasting statistics for decision making

"RFID demands a measured response."[1]

Oliver Hedgepeth

Forecasting is a process for predicting a future event. Forecasting is used in every aspect of business decisions, from production scheduling, inventory trends, personnel, and all manner of facilities management. RFID is no exception to forecasting demands. The use of different RFID metrics dictates that trend analyses are needed to help control and understand the potential return on investments.

In the 1960s, Switzerland controlled nearly 70% of the world watch production market. It had a history of steady market share growth for decades. In the 1960s, the forecast for the market in the next 20 years should have been more. However, by the 1980s, the Swiss market share was dropping significantly. The winner was Japan with the advent of its new electronic quartz technology capturing nearly 40% of the new market, although Switzerland was the innovator in such technology.

In the 1980s, Symbolic was the leading contender to lead the computer industry in its artificial intelligence or knowledge engineering (KE) computers, being used by military and industry to develop smart software for logistics and maintenance needs. Its hardware and software was significantly faster and more sophisticated than Sun or IBM or any other computer at that time, at least in the KE world. During the mid-1980s to 1990s, the growth of military and commercial expert systems dominated the field. Ten years later, no one could find Symbolic computer systems; Sun and others had pushed them out of the business.

Technology forecasting can be economic, technological, or demand. For economic forecasting, you can address the business cycles, such as inflation rates, money supply, and other traditional financial metrics. For technological forecasting, you predict technological changes and new product sales. For demand forecasting, you predict how existing product sales will be changing in the future. The focus in this section will be on technological forecasting for RFID. The other forecasting methods lend no special extra tools to help understand the RFID metrics needed to succeed in your business decisions.

One of the key parameters in forecasting is time. In fact, time is a dimension of RFID metrics that we discuss often in this book. The timeframe is typically short-range, medium-range, and long-range. We will define short-range as up to 1 year, but generally more than 3 months. This is useful for pilot RFID tests. Medium-range will be a time span from 1 to 3 years. Long-range will be for 3 to 5 years. It is interesting that some of the RFID plans that have been openly discussed contain plans to incrementally test and expand RFID implementation in the medium- to long-range, 3 to 5 years.

Forecasting for RFID is similar to other technological forecasting methods. You first have to determine the purpose of your forecast. This could be to increase profits, to increase market share, to maintain market share, or to keep from going out of business. With RFID, it seems that all these might seem reasonable purposes.

The second step is to establish a time horizon. For RFID, what should this be? Given the mandate from Wal-Mart and the Department of Defense, the suppliers to those two organizations are already in the process of investing or investigating using passive RFID tags on pallets, containers, and special items, and for security. The time horizon has to be reasonable within your ability to gauge the system's impact to your organization. Do you consider your organization a closed- or open-loop system?

The third step is selecting the proper forecasting technique. For RFID, there are two kinds: One is quantitative and the other is qualitative, based on similar experiences we had when forecasting the use and cost of expert systems or knowledge systems for logistics operations in the 1980s.

The fourth step is gathering data. The assembling of data and examining and analyzing patterns is more problematic today with RFID given the assumptions you bring to this analysis.

The fifth step is preparing the forecast for presentation to some decision maker. This is a crucial step in any forecast, but is made especially difficult for RFID applications; again, the lessons learned from the expert systems and knowledge engineering days of the 1980s point to pitfalls of not paying attention to certain lessons learned.

The sixth step is monitoring the RFID forecast. This is not like monitoring your inventory levels or production scrap pile levels. This is totally different, given the cultural and political aspects of RFID applications that could change your assumptions and the vulnerabilities of those assumptions.

For RFID forecasts, quantitative methods will be useful when the metric used in collecting stable and historical data is accurate and dependable, just as in any logistics or supply chain database. This will use traditional and descriptive statistical analysis methods.

For RFID forecasts, qualitative methods will also be very useful, because it is possible that other data are not easily quantifiable, when the situation is vague and little data exist, which is the case for RFID. This is usually the case for new products and new technology. Although RFID is not new, the small size and cheap price bring passive RFID technology into this qualitative realm. Whenever there is a new technology insertion, you will need to rely on judgmental forecasts until real data can become available; for RFID, it is possible that you will not be able to abandon the qualitative forecast until about 2010. Associative or causal methods are not necessarily more accurate than time series, but for RFID they need to be used in mental model analysis to balance the scarcity of data.

Metrics for an RFID pilot program

An objective of this book is to provide a basic analysis resolution of alternative RFID information management processes, which could affect future company or government policies, procedures, standards, software, and tracking and tracing technology. It identifies the direct and indirect advantages and disadvantages of RFID systems for record keeping. This description focuses on two interrelated factors: (1) the extent to which RFID systems aided in fulfilling the organization's inventory management regulations and policies, and (2) user perceptions with regard to the value and usefulness of RFID technology compared to current bar code and manual tracking processes.

Problem statement

Any evaluation will provide data collection and analysis services to the organization to determine the results of the RFID pilot project that support the scope and objectives. This evaluation includes the development of a methodology for this process that addresses project needs for information collection instruments. An approved data collection methodology must be developed.

A first step in defining the analysis problem to be addressed is to understand the environment of the organization's current logistics requisition and inventory record-keeping requirements. This involves developing a visual image of how the current organization's requisitioning process is conducted in a manual and automated manner. Within this context, five questions and related hypotheses form the problem statements for this evaluation, as shown below:

Problem Statement 1: How are the capture and filling accuracy rates affected by the introduction of an RFID tracking system? The related hypotheses are as follows.

A. Hypothesis 1 (H_1): The current manual/bar code tracking accuracy (MTA) rate is equal to the RFID tracking accuracy (RTA) rate.

H_1: MTA = RTA

This metric will show how the accuracy rate of each requisition for items or products (by requisition document number or some other label) by individual commercial or military units or organizations equates to the accuracy of the same requisition method using RFID technology over time. The metric will be in numbers of requisitions that are classified as not accurate. The goal of this hypothesis is to reach a point where the numbers of missed classifications are equal with both inventory tracking methodologies. If a difference between the two metrics, MTA and RTA, is greater than 10%, we can say that there is a significant difference. If the difference is less than 10%, it is not significant. For example, if MTA is 350 requisitions identified with 35 inaccuracies, and the RTA shows 633 requisitions with 33 inaccuracies, then we could show the following:

Given this information, we could conclude that this hypothesis shows that MTA RTA as an absolute number (that is, $35 \neq 33$). But does this inequality hold for percentage differences? The answer may be no. This is due to the number of inaccurate differences being 2, or 6% (100% – 94%) in this example, which is less than 10%. We could conclude from the percentage differences that H_1 is true and that MTA = RTA.

Although the total number of requisitions captured shows a 55% difference, this number is only useful for administrative purposes at this time. The difference in number of accurate requisitions is meaningless in this hypothesis. However, the overall numbers of requisitions that are inaccurate still need to be addressed so as to make this number approach 0%.

B. Hypothesis 2 (H_2): The manually or bar-coded processed requisition accuracy rate demographic (MRAD) for two-level factors are the same.

H_2: $MRAD_1 = MRAD_2$

The metric used for this hypothesis is the demographic designation for the different types of persons or units manually processing or filling requisitions. For example, the number of requisitions being filled by two different units or organizations are one group, $MRAD_1$, which could be significantly different (in accuracy) from a second unit or organization at a different customer location, $MRAD_2$. The two-level factor would then be accuracy rates of different types of users or customers. Similar demographic information at two levels could possibly be used, such as class of item, perhaps food or textiles. Absolute numbers and percentage differences as shown for solving Problem Statement 1 would be used.

Problem Statement 2: What is the probability of finding an accurate match using either a manual or bar code method compared to the RFID method? The related hypotheses are as follows:

C. Hypothesis 3 (H_3): The probability of manual or bar code controlled requisition filing (MF) is equal to the probability of RFID requisition filing (RF):

H_3: P(MF) = P(RF)

Inaccurate data can cause the tracking system to be inefficient. The metric for testing this hypothesis is based on an assessment of a baseline of requisition records filed accurately over a period of time by a set number of users at various nodes along the supply chain and by RFID. The probability of finding either a correctly or incorrectly classified requisition record is defined as a Bernoulli process, which follows a binomial distribution. This probability will be defined on a monthly (or daily) basis and evaluated constantly during the test period. The final results of this analysis will provide a probability statement with a specified and measured level of confidence.

Problem Statement 3: What is the rate of learning with the users of the manual or bar code system compared to the RFID system?

D. Hypothesis 4 (H_4): The learning rate of users of the manual or bar code system (the change in accuracy rate) is directly proportional to performance increases.

H_4: $P(MF_{n-1}) = P(MF_n)$

E. Hypothesis 5 (H_5): The learning rate of the RFID system (the change in accuracy rate) is directly proportional to performance increases.

H_5: $P(RF_{n-1}) = P(RF_n)$

Learning only occurs if the users of the manual or bar code system and the RFID system improve their accuracy rates. From the probability established from Problem Statement 2, we then use this probability as a measure of both the human accuracy rates and the RFID accuracy rates. By comparing the probabilities and the accuracy count that is used to measure the probability, we can find any change from month to month or, if finer-grain measures are needed, from day to day. Thus, $P(MF_{n-1})$ equates to calculating the probability for the base on the initial month, such as January (which equates to n = 1, and then to n – 1 = 1 – 1 = 0), whereas $P(MF_n)$ would be the probability for the next month, February (which equates to n = 2). Therefore, the comparison of learning could not be calculated until we received data for 2 consecutive months. Then, $P(MF_{n-1}) = P(MF_n)$ would be $P(MF_{2-1}) = P(MF_2)$ or $P(MF_1) = P(MF_2)$.

The expected, repeated calculations of this probability are due to the feedback mechanism from the experts to the current users of manual or bar code processing of requisitions filed or processed inaccurately, with expected constant learning, which will bias previous results taken prior to any feedback point. Without this feedback point, a reduction in the frequency of checks will not be possible and the probability would not be subject to change, which reflects learning on the part of these people or units.

Problem Statement 4: What is the probability of customer wait time (CWT) of a manual or bar code system compared to the CWT of the RFID system?

F. Hypothesis 6 (H_6): The probability of a manual or bar code system of processing requisitions P(MCWT) is equal to the probability of using RFID, P(RCWT).

H_6: $P(MCWT) = P(RCWT)$

Problem Statement 5: What is the probability of requisition wait time (RWT) of a manual or bar code system (MRWT) compared to the RWT of the RFID system (RRWT)?

G. Hypothesis 7 (H₇): The probability of a manual or bar code system of processing requisitions P(MRWT) is equal to the probability of using RFID, P(RRWT).

H_7: P(MRWT) = P(RRWT)

The metric used for Hypothesis 6 and Hypothesis 7 is based on an assessment of a baseline of current industry or military organizations' supply requisition actions compared to the use of RFID technology over the study period of time. The probability of finding a higher percentage from the manual or bar code system compared to the RFID system should follow a binomial distribution. The probability will be defined on a daily and monthly basis and evaluated continuously during the test period. The final results should provide a probability statement with a specified and measured level of confidence for CWT and RWT.

Scope

An indirect result from solving these problems is the description of the workflow of documents and tasks in the use of the organization's supply chain tracking and tracing process. This process may prove to be a sequential or parallel process or work activities or a complex set of processes. The process could include a decision-chain process, which relies on the organization's supply milestones and decision points to produce the final supply tracking product. Conversely, it could be an event-driven process based on a chain of command or chain of custody or manual or bar code and RFID data filing events.

The results of this kind of effort will identify changes over time among the filing methodologies for the organization's supply requisitions. Although the result of any pilot study may not completely answer any organization's business process re-engineering questions of workflow, it may offer suggestions toward minimizing customer wait time, requisition wait time, and other factors of business customer or war fighter confidence in the supply system.

Assumptions

There are several assumptions for statistical analysis of the data that could be produced from the target records provided by any organization under study.

- Records and files are captured with a normal probability distribution.
- Errors in one database are independent of errors in another database.
- Files are not damaged or altered prior to study team having accessibility.
- The data capture technology for RFID records does not fail and event records are uniquely identified.
- The probability of each RFID or other file record being correct in any database is the same.
- The organizational expert does not make mistakes when checking or providing the study team databases.

These assumptions provide a basis for measuring the problem statements and hypotheses. But some of these assumptions could be violated, resulting in an evaluation problem or issue. If so, then the problem statement and hypothesis will need to be revisited.

Research design and methodology

Research design is a plan for conceptualizing the structure of the different variables that are studied as RFID technology is evaluated for pilot study use. It implies how a research situation can be controlled, and how the organization's data are to be measured and analyzed. The research methodology indicates how to take observations in a systematic and standard manner. This research evaluates and compares the capture and accuracy rates among the current manual or bar code system and the test technology of RFID being examined, and tracks the trends over time. This evaluation requires specification of the data models, the data stratification strategy for the building and testing of these data models, and logic for assessing performance of these models. The research design and methodology should be divided into five areas:

- Data to measure
- Current system evaluation and data model
- Random samples
- Data collection (quantitative)
- Data collection (qualitative)

Data to measure

The scientific measurement activities for any RFID research are the systems intersection of the problem statement, hypotheses, data types, and data collection instruments. This is shown in Figure 7.1

The data are of three types. First, data are collected as a single point in time (e.g., for a month). The objective is to examine such data and then to extrapolate any revealed trends or relationships to a larger population. Figure 7.2 shows a notional example of data that might be collected in the month of January for six test groups of supply items. The chart indicates daily averages of records captured at some supply node for January for each of the seven Test Nodes (TNs). Each Test Node is composed of various numbers of people, processes, and procedures. The implication is that each TN captures a different number of requisition records. Further breakdown, by node location or demographics for supplier or customer, could show possibly underlying trends. The implication could be that each TN captures a different number of requisitions of file entries. Further breakdown by demographics could possibly show other underlying trends.

Problem Statement	Hypotheses	Data Types	Data Collection
Problem Statement 1: How are the capture and filling accuracy rates affected by the introduction of RFID tracking system?	**Hypothesis One (H_1):** The current manual/barcode tracking accuracy (MTA) rate is equal to the RFID tracking accuracy (RTA) rate. **H_1: MTA = RTA** **Hypothesis Two (H_2):** The manually or barcode processed requisition accuracy rate demographic (MRAD) for two-level factors are the same. **H_2: MRAD$_1$ = MRAD$_2$**	Count of records at a point in time (e.g., hour, week) at regular intervals (e.g., every Friday, 5:00 PM Alaska time) Check accuracy of records and files. Demographics of requisition and customers.	Electronic query for record counts. Physical comparison. Interview or survey report for demographic information.
Problem Statement 2: What is the probability of finding an accurate match using both manual or barcode method compared to RFID method?	**Hypothesis Three (H_3):** The probability of manual or barcode controlled requisition filing (MF) is equal to the probability of RFID (RF)requisition filing: **H_3: P(MF) = P(RF)**	Changes in capture and accuracy rates over time in weeks and months.	Electronic query for record counts. Capture and accuracy rates.
Problem Statement 3: What is the rate of learning with the users of the manual or barcode system compared to the RFID system?	**Hypothesis Four (H_4):** The learning rate of users of the manual or barcode system (the change in accuracy rate) is directly proportional to performance increases. **H_4: P(MFn$_{-1}$) = P(MF$_n$)** **Hypothesis Five (H_5):** The learning rate of RFID system (the change in accuracy rate) is directly proportional to performances increases. **H_5: P(RFn$_{-1}$) = P(RF$_n$)**	Changes in capture and accuracy rates over time in weeks and months.	Electronic query for record counts. Capture and accuracy rates

Figure 7.1 Data sources and collection methods.

A second type of data is aggregate and will be collected at monthly intervals; expect the data would be collected for just one TN over time. Such a time series could indicate a period of record or filing losses for some gap in time, which could indicate a pattern. Such trends over time are important in forecasting supply requisition filing accuracy, because any analysis of a point in time does not provide a complete, potential prediction of future events.

Also, because causal conditions seldom remain constant, we need to add the qualitative evaluation factors to these quantitative data displays. This will aid our understanding of what may be the underlying causal factors for the changes.

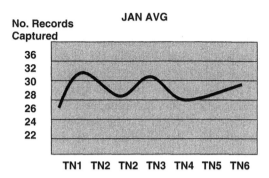

Figure 7.2 Notional example of data that might be collected in the month of January for seven test groups of supply items.

The third data group is from any survey or interview using formal questionnaires, interviews, and focus group sessions. This type of data gathering yields the greatest amount of information if the respondents are willing to be open and to spend time answering questions. Using in-depth interviews should be sufficient to cover the range of potential problem areas. For example, Figure 7.3 may help to identify confusion on RFID or manual and bar code record keeping or tracking protocols to follow, as a causal factor. Also, some respondents could be shown to not be at work due to holiday vacation for January, as shown in Figure 7.3.

As can be seen from Figure 7.3, the results from the qualitative data collection tell one story of positive improvement, while examining the raw time series data may create an opposite picture. Only when the data from all sources are combined does the complete story emerge. If the results from Figure 7.3 have been all negative themes as well as metaphors from the respondents, then the results from the time series data would be a stronger case for calling a management decision with the organization.

January	February	March	April	May	June
90% work force on vacation for two weeks.	People seem excited about the RFID program	Loss of three people due to rotation	Loss of test node due to security reasons	Loss of people to rotation assignments	Loss of large number of people due to rotation
Change in DOD policy	Focus groups produce positive themes	Computer software bugs causes loss of confidence	Confidence soars over use of RFID software fixes	Certain test sites see themselves as RFID pioneers for DOD	Test sites requests changes to procedures for manual and barcode records capture
Change in DOD computer system	Workloads shift for RFID test groups	Shift in software for RFID capture loses time to capture files	DOD people more excited than ever about RFID system	Focus groups uncover more positive themes on using RFID	
Confusion about DOD and Industry RFID pilot study					

Figure 7.3 Notional data from surveys and interviews.

Current system evaluation and data model

Figure 7.4 illustrates the process for evaluating the RFID test system and corrective action loop. The evaluation aspect of this system is designed to be applied on a continuing basis to provide system measures at designated points throughout the life cycle of the RFID test period. Problem areas are identified in the evaluation and are reviewed in terms of feasibility for corrective action. Corrective action is to be accomplished in response to the RFID system deficiency (i.e., the RFID hardware and software fails to identify and classify supply item records or file items) or human error (e.g., misunderstanding of how to handle RFID equipment).

The RFID box is where the human decisions and RFID methodologies occur. The survey is where the qualitative aspects of this study are determined. Both of these process steps will continue throughout the life of the analysis. The data collection is drawn from the daily, weekly, or monthly supply record activities and any scheduled surveys. The raw data is kept in original electronic form as automated supply record capture information. The data analysis should use a series of analysis models, such as frequency distribution, graphing, measure of central tendency and variance, measure of relations, and regression analysis. The regression analysis could be both traditional mathematical models and neural network models in the case of missing or questionable data sources. These analysis models could be used to evaluate the RFID system. If any problems are identified within the RFID system, which includes RFID computer interface software and human decision making, the problem is identified and evaluated with the appropriate analytical tools. In fact, it is most likely that problems could be identified from statistical analysis rather than from the supply chain users from the organization or vendors. Problems are evaluated using the data bank of coded and processed data that represents the time history of any RFID analysis.

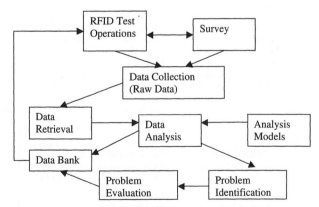

Figure 7.4 The process for evaluating the RFID test system and corrective action loop.

Statistical analysis

At the heart of any research study for RFID use is the methodology used to collect responses from individual supply chain nodes and individuals who are working at those nodes to help collect the raw data stream of requisition files. The best case for data collection would be to consider that during this test, the total number of requisitions could be collected during the complete data stream on a daily basis for 6 months. However, traditional data streams may be more than a year old for analysis purposes, and it is not certain that all the data files will be available. Therefore, a random sampling may be needed, and provisions for random sampling need to be considered.

When data are obtained, there is always the chance that the estimates might differ from the true figures that could be obtained from the entire target population. This difference is referred to as *sampling error*. There are several useful rules that can be used to interpret the accuracy of the data, which is done through statistical analysis.

Capture data analysis

To evaluate the quality of data being produced, it is necessary to analyze the accuracy of the data capturing methods. Based on estimated rates of

Jan	TN1	TN2	TN3	TN4	TN5	TN6	TN7	TN8
1	58	43	6	59	39	43	21	59
2	39	59	36	53	42	19	51	56
3	56	44	44	38	13	28	47	14
4	33	36	38	59	2	53	25	2
5	26	24	18	32	36	11	58	12
6	54	47	54	4	18	5	35	34
7	33	38	24	8	42	23	31	60
8	39	55	27	12	14	57	1	38
9	24	19	2	25	33	30	40	51
10	54	59	50	13	54	6	11	30
11	13	8	21	39	54	41	49	40
12	37	48	39	49	0	57	35	22
13	19	49	31	49	39	7	4	28
14	0	25	56	41	52	22	27	32
15	30	5	44	50	18	7	34	25
16	27	40	35	30	46	9	15	39
17	52	55	8	51	35	18	46	10
18	43	32	8	20	56	5	2	29
19	59	37	44	26	26	59	38	45
20	52	54	27	3	27	44	29	55
21	22	54	14	50	49	9	8	6
22	10	11	28	8	34	8	2	52
Total	780	842	654	719	729	561	609	739
M	35	38	30	33	33	26	28	34
SD	17	17	16	19	17	19	18	18

Figure 7.5 Notional capture data for eight supply chain nodes.

requisition filing activity, there is a potential for several thousands of records over a month or over the entire test period. A sample data set for capturing requisition data at various test nodes (TNs) along the National Defense Department supply chain is shown in Figure 7.5. Basic statistics such as the mean (M) and standard deviation (SD) are calculated.

Reference

1. Hedgepeth, Oliver, RFID demands a measured response, *RFID Journal*, April 10, 2006, accessed by Internet May 1, 2006.

appendix A

Study questions

The following assignments reinforce the learning objectives in the chapters of this book. The questions can be assigned to the student depending on whether the course being taught is in a traditional brick-and-mortar institution or as part of an e-learning course offered via the Internet. These questions will require the student to go to the Internet or library and search for the answers, which may not be given in this book.

1. Go to a local or university bookstore and speak to the manager about how they currently use bar codes for keeping track of and tracing their books. Discuss the possible impacts of RFID technology. If they are currently using RFID, try to find the problems they are solving with RFID. If they are not using RFID, explain the technology and see if they can define a possible problem that can be solved using RFID technology. Once you have your problem defined, with the help of the store manager, explain why the problem is not complete, and why the problem may not be the real problem to be solved.

2. Find a municipal parking garage or system or one on a university campus and speak to the manager about how they currently use code numbers on parking permits or bar codes for keeping track of cars parked in the parking lot. Discuss the possible impacts of RFID technology. If they are currently using RFID, try to find the problem they are solving with RFID. If they are not using RFID, explain the technology and see if they can define a possible problem that can be solved using RFID technology. Once you have your problem defined, with the help of the parking manager, explain why the problem is not complete, and why the problem may not be the real problem to be solved.

3. Visit a fire department and speak to the chief or captain about how they currently use code numbers or bar codes for keeping track of fire equipment. Discuss the possible impacts of RFID technology. If they are currently using RFID, try to find the problem they are solving with RFID. If they are not using RFID, explain the technology and see if they can define a possible problem that can be solved using

RFID technology. Once you have your problem defined, with the help of the chief or captain, explain why the problem is not complete, and why the problem may not be the real problem to be solved.

4. What would you advise on how to produce a return on investment (ROI) for RFID implementation?

5. What would you advise about sharing results of passive RFID pilot tests?

6. Is the license plate data from the Smart Passport meaningless to anyone if they do not have access to a database file on the person being scanned? What would you advise U.S. citizens or DoD shippers and logisticians or even soldiers who use Smart Passports, when they are traveling outside of the United States?

7. What would you advise about the improved visibility of material shipped to the soldier?

8. What would you advise management in order to make sense in distilling and mining the large volume of logistics data coming from the RFID system?

9. How would you handle the workers' concerns for wanting to call someone about a backlog, the flight schedules, the pallets, or the nets that might be missing?

10. What would you suggest to management about the continuous stream of data coming from the inventory count?

11. What would you suggest to management about balancing the technology, business process, and control systems of the business since RFID was implemented?

12. What would you advise in adding RFID data collection systems to replace the workers?

13. What would you suggest to the manager to ensure that their monthly lumberyard inventory count was accurate? Would you suggest continuing with the same manual method or use bar codes or RFID?

14. How many passive RFID tags would you suggest be used by management to track their pallets and cases?

15. How would you suggest to the manager different ways to handle congestion that seems to be coming from the volume of distribution data?

16. What would you suggest to the manager about possible reductions in the personnel needed for their logistics operations?

17. Go to the Internet and find 10 different RFID applications. List the problems they think they are solving and why. Do you think they have thought through the problem sufficiently? Explain your answer.

18. Go to the Internet and find 10 similar RFID applications for use in manufacturing and materials management. Examine the savings that RFID technology is providing to the manufacturing organization. Is there a trend in the type of savings that you can find or that you assume from your readings? Explain your answer.

19. Find 10 different applications, one each in a different country. How is each application being measured for success to the organization?

Define, explain, and compare each of these 10 (or however many you find) metrics. What trends did you find? Why?

20. Examine one application of RFID in transportation of goods or cargo that can be explained with the 10 metrics provided above. Which metric of the 10 appears to be dominating to providing a better understanding of the application usefulness? Explain your choice. Then, identify which metric of the 10 appears to be less effective than the others in understanding the application usefulness. Explain your choice. Why are these two metrics so far apart from each other in usefulness?

21. Using the transportation concepts and perspectives presented in this book, take a product from a real company that is using RFID technology and map its movement network. Contact a real company for this assignment. Explain that you want to trace the routes of just one type of product from origin to final destination. Examine the network in terms of these concepts and perspectives and give an explanation for the pattern you see emerging from this network. How does RFID technology explain these concepts differently than if RFID had not been used? Discuss the results with the manager of the company. What are their opinions?

22. If you can, go to a seafood processor and ask how it might use RFID technology to track its seafood product, fresh, frozen, or canned. This can be a processor in the East, West, South, or even in Alaska, or in a different country such as Chile or Japan. Describe how RFID technology could help solve their problems.

23. What other factors besides those listed in this book might be affecting performance of the use of RFID tags in supply chains, logistics, or transportation activities? Evaluate each of the factors presented and try to find reasons that would not be a good factor for measuring performance. Of these performance factors, which is the most important? Which is the least? Why?

24. Visit a warehouse or a business with a loading dock. Examine the entrance where the products or goods enter the building, and determine whether or not a forklift is used. Examine the possible places for antenna placement. Take pictures and draw a diagram of where antennas should be placed to read passive RFID tags that have a read distance of 8 feet from the antenna. What issues do you see in the placement of such antennas in this environment?

25. Visit a plant that makes milk or beer or cola beverages and examine its operations for moving its product. What kinds of containers are the basic items, such as milk cartons, bottles, or cans, placed into? For example, milk might be placed into plastic box-like crates that hold only four gallons each. Then, each of these crates might then be stacked into a pallet of any number of these smaller crates. Examine the possible placement of RFID tags on these initial crates that hold the beverage. What is the issue in being able to effectively use tags

on such packaging material? Look at the whole supply chain in your answer, from the time the crates are filled and leave the plant to the time the empty crates come back to the factory, to being refilled again with product. What problem do you see in this process that would make use of RFID tags not work?

26. In the previous two problems above, document the environment around the use of the RFID tags. Is there dust, rain, or natural causes that could prevent the use of RFID tags effectively? If so, how would you propose getting around these environmental obstacles?

27. Visit a grocery store and pick an aisle of pharmacy goods. Look at a shelf at eye level. Write down all the product names on the shelf. Examine the containers. How many different bar code sizes are displayed on the package or item? How many items are on the shelf in total? If you can, take a photo of each item and then make a recommendation on how you would place a passive RFID tag that was 1-inch-square on this package, given that that would be the only size RFID tag you were allowed to use. What issues have you found? What problems have you found? What solutions would you propose to be able to use an RFID tag on each of these products?

28. What business case can you determine where RFID tags are not needed now and in the foreseeable future? A hint is to find some product that currently does not use bar codes. You may have to go into the materials management side of business to find this answer. Document why you believe that this product or organization or company would never use RFID technology.

29. Search the Internet for the best business cases for use of RFID technology. When you find these business cases, describe what they have in common. What seems lacking in their description of the use of the technology?

30. List 10 reasons for why RFID will not work. Do not use any from the list that is given in this book. Give your source of information. Do you agree or disagree with these reasons? Why or why not?

31. Find an example of an organized anti-RFID group that wants to stop the spread of RFID technology. What are their assumptions about RFID technology and its impact on society and business? Do you agree or disagree with the anti-RFID group? Why or why not?

32. Examine one company's RFID applications from the retail side of business and develop and define the paradigm of RFID as different from that of the company using bar codes. Explain your paradigm in terms presented in this chapter.

33. Find a company, such as a lumberyard, that does not use bar codes. Find out how it conducts its inventory management. What is the paradigm that this company uses and how might bar codes help change this paradigm? Can its paradigm be changed? Can RFID help shift the current paradigm without the company ever using bar codes first before using RFID?

34. Visit a grocery store and talk to the manager of meats, fish, or produce. Determine the range of different temperatures needed for all perishable products within the store, and in the storage facilities. If the store manager will let you, try to spend some time in the loading area when fresh and frozen goods are being shipped into the store. What procedures do you notice? What is the range of time that perishable inventory is kept in the local warehouse or storage facility on the premises, if any? Determine what temperature control processes and procedures influence decisions made about purchase, retention, and finally waste disposal of perishable food items under temperature control. Obtain any written policy statements provided to the employees about the handling of fresh and frozen food items. Following this, go on-line and research the Food and Drug Administration and Department of Agriculture rules and policies for the use and handling of perishable food items.

35. Choose four countries and check on the legislative issues and laws governing the handling, storage, and transportation of fresh and frozen goods. Pick the United States as one country. Then choose one country from Europe, one from South America, and one from Asia. Contrast and compare your findings. What patterns have you found? What issues did you find? What surprised you the most about what you found?

36. Consider a logistics or supply chain system that is dedicated (according to the newspapers or trade magazines) to implementing passive RFID tags. How would you go about forecasting why this company should invest in RFID technology, and what factors would you measure to help in creating a forecast for the demand for using RFID?

37. Consider a logistics or supply chain system that is dedicated (according to the newspapers or trade magazines) to not implementing passive RFID tags. How would you go about forecasting why this company should invest in RFID technology, and what factors would you measure to help in creating a forecast for the demand for using RFID?

38. You are the CEO of a milk-producing company in Alaska. Your customers are only those people who live in Alaska as well as the five military bases. You are not sure if you should invest in passive RFID technology, even though your customers are Wal-Mart and the military. What are the advantages of using the following types of people to help you make a decision?
   A committee of logistics professors from the local university
   A committee from the city's economic development commission
   A committee from the company's union workers and managers
   A committee of paid consultants from MIT who are experts on RFID

39. What are the factors that you would use to make a decision to replace bar codes with passive RFID tags?

40. Visit a security firm or your local police station, as well as two businesses, and talk to someone about how they see using passive RFID

technology for individuals to use to gain access to a building. Discuss what different technologies are used besides passive technologies, such a closed-circuit television, biometrics, face-to-face discussion, voice recognition, fingerprints, and any other methods. Identify their policies for use of such personal tracking and tracing technology.

41. Contact and visit a public library that is using RFID technology. Are they using the RFID technology as a remote sensor with book drops, or as readers when checking books in or out of the public patron desk of the library? Discuss with the library's person in charge of security or inventory exactly how RFID technology has impacted their employees' work, patterns of behavior, and accuracy of shelf stock inventory. How has theft decreased (or has it) since the inception of RFID tags on library items, books, CDs, DVDs, and other artifacts the library lends to the public? Does the library use handheld RFID readers to scan the bookshelves for proper placement of books? How much time does the staff save on reshelving books due to the use of RFID? (Note: If you cannot visit a library in person, then contact a library by phone to ask these questions; then talk to the local library about RFID technology after your interview with a library that does use RFID. Obtain their views of the answers to the questions you have found from talking to the library that uses RFID.)

42. Check legislation on the use of passive RFID tags for use in libraries in your home state, as well as such policy actions within the city or county where you currently reside. Are there any restrictions of the use of RFID tags by librarians or libraries as this technology comes into contact with the public? Check with information published about the library board of directors, or whatever governance body acts to oversee the activities of your local library. Talk to the libraries and their administrators in your state about the public policy, privacy, and security aspects of the use of technology in their library.

43. Contact at least four companies in the private sector that use RFID access cards. Explain that you are conducting a class project on the policies used by different companies for this RFID technology. You may decide what questions to ask. For a start, however, do ask the following questions: How long has the company been using RFID cards for employee access? Where are these cards used in the company? What type of information or data is collected from these cards? How long is this access information kept in the company databases? And remember to think of the what, when, why, where, and how aspects related to the use of RFID tags for access to the company. From these questions, determine what, if any, common principles seem to underlie all four companies' policies for use of data generated with RFID access cards. Also, try to determine if these policies are communicated to the employees.

44. Contact a large global company that uses RFID access cards. Ask about the policies for use of these access cards by employees in the

United States and then how these policies differ in the different countries around the world where this company operates. Depending on which company you choose, you could be looking at upwards of 30 different countries. You already know that policies for the use of personal information are different in some countries. So, once you find these different countries for this company, explore a little more on those countries' policies toward the use of RFID technology. Compare and contrast this country's RFID access card use policy to that of these foreign countries. If you are taking this course outside the United States, then start with a company that is headquartered in your country and follow the same questioning process.

45. This is a problem to test your ability to think about probabilities and how they are used. An Internet reader survey was conducted in 1996. The purpose of the reader survey was to explore trends in attitudes about local newspapers concerning their credibility.

Data was gathered from people in Petersburg, Richmond, and Chester, VA. The basic question asked was whether they read local, daily newspapers, and if so, how often. Only people who used Internet news at least 4 days a week were interviewed. The data gathered is shown in Table A.1 and Table A.2.

Questions:

From inspection of Table A.1, what are your findings of each of the six characteristics?

What factors (from inspection of Table A.1) could have influenced (or probably did influence) your findings, as a positive or negative factor, or both?

From inspection of Table A.2, what are your findings of each of the six types of questions used in the survey?

What factors (from inspection of Table A.2) could have or probably did influence your findings, as a positive or negative factor, or both?

What conclusion can you draw from your inspection of Table A.1 and Table A.2 regarding the performance of the news media?

What technological trend(s) could you forecast from your conclusions?

What should be added or deleted from the technical data? Why?

The use of statistical analyses is optional. The length of your answers is optional. However, the purpose of this exercise is to think about the concept of forecasting, and those relevant items that can be analyzed from the data presented. Use the language of technological/technology forecasting. With a thorough inspection of data, your assumptions, and the review of the number of variables that deserve attention, four to five pages would be sufficient to answer all seven questions. But, if you need more space, go for it. Format is a narrative story—to evolve and grow from your inspection of the data—not just a list of bullets or list of items.

Table A.1 **Internet survey results**

I. Demographic Characteristics				
Characteristics of Data Gathered from Readers' Survey	Petersburg	Richmond	Chester	All
Sex				
Male	5	5	5	15
Female	5	5	5	15
Race				
Majority	8	8	8	24
Minority	2	2	2	6
Age				
Range	29–72	23–66	24–73	23–73
Mean	47	44.1	46.1	45.7
Education				
College graduate	4	6	5	15
Some college	2	3	2	7
High school graduate	3	1	3	7
Less than high school	0	0	0	0
No response	1	0	0	1
How closely follow the news				
Very closely	5	4	2	11
Fairly closely	4	6	8	18
Not very closely	0	0	0	0
Not at all closely	0	0	0	0
No response	1	0	0	1
Political views				
Strongly conservative	2	1	0	3
Conservative	1	3	5	9
Moderate/middle road/mixed	6	4	4	14
Liberal	0	2	1	3
Strongly liberal	0	0	0	0
No response	1	0	0	1

Table A.2 **Internet survey results**

II. Media Usage

Questions to Readers of the Survey	Petersburg	Richmond	Chester	All
Most important news source				
Local TV news	7	1	3	11
Network TV news	3	2	0	5
Local newspapers	6	7	5	18
National newspapers	0	1	0	1
Radio news	0	1	2	3
Some other source	0	0	0	0
Days of the week read local paper				
Range	4–7	5–7	2–7	2–7
Mean	6.5	6.8	4.9	6.1
Quality of reporting in local paper				
A (Excellent)	0	0	1	1
B (Good)	9	9	8	26
C (Fair)	1	0	1	1
D (Poor)	0	1	0	1
F (Failing)	0	0	0	0
Mean (A = 4, F = 0)	2.9	2.8	3	2.9
Change on reporting quality in local paper				
Gotten better	3	3	4	10
Gotten worse	1	0	1	2
About the same	6	7	5	18
Percentage believed in local paper				
Range	70–90%	5–95%	60–95%	5–95%
Mean	80.00%	74.80%	77.00%	77.20%
Change in amount believed in local paper				
Believe more	1	1	3	5
Believe less	1	4	1	6
Believe about the same	8	5	6	19

46. Examine the following chart (Table A.3) of assumptions for the major airports in the United States. Where would you see that RFID could impact the five areas of assumption-based planning? What changes would you make to this chart given the assumption that RFID is to become as commonplace as bar codes for transporting luggage, passenger tickets, and air freight?

Table A.3 US airport analysis of assumption affecting future airport growth

Assumptions

	6% Growth	Asia Largest Shipper of Products	Great Circle Route	Cargo Transfer Authority	Globalization of Markets	Liberalization of Markets and Air Service
VULNERABILITIES						
World Economic Recession	X	X			X	
World Political Relationships	X	X		X	X	X
Larger Airplanes and Overflights	X		X	X		
Competitive Expansion or Loss of Cargo Transfer Authority	X			X		X
Measurement of Globalization	X				X	
Restrictive Air Service Policy	X			X	X	X
SIGNPOSTS						
Static Growth in Transfer Activity	X	X			X	
War in Iraq	X	X			X	
CRAF Activation	X			X		
SHAPING						
Diversify Markets	X			X	X	X
Increase Marketing Efforts	X	X		X	X	X
HEDGING						
Expansion of Cargo Activity		X	X	X	X	X
Logistics Hub		X	X	X	X	X

appendix B

Case studies

These case studies are designed to promote critical thinking of the topics within this text. The cases are created from real examples, some fiction added, to help the reader in solving similar real-world problems in implementing RFID technology.

These case studies are designed to be answered by undergraduates as well as graduate students. The graduate student is expected to bring a little more statistical analysis rigor to analyzing the case problem or problems.

These cases are in addition to and separate from the study questions in Appendix A. Also, as the student or reader progresses through the chapters and lectures, so will the answers to these case studies be expected, in some cases, to pull from the previous chapter's study. These cases will challenge some preconceived ideas about technology insertion, specifically, the implementation of RFID technology to supply chain demands.

The author welcomes feedback and can be contacted at the University of Alaska Anchorage web site, at afwoh@cbpp.uaa.alaska.edu, or through the publisher of this book.

Case study 1

Smart pallets for Harman's Repair Station, Inc.

Barry Benton walked into Harman's Repair Station a very excited man. He had just completed a class in how to use RFID for supply chains and he thought that RFID would be a perfect fit to save time and money at their igloo repair shop. All he had to do was convince the president, Don Harman, that RFID was something he should jump on right now.

Current situation

Today was Saturday, and Barry was the first to arrive, open the door to the repair warehouse, and start the coffee. Today was to be a slow day, one filled with paperwork from last week's netting, pallets, and igloos for luggage storage repaired from several Air Force and commercial cargo planes.

Barry walked around the warehouse inspecting everything to make sure the week's workspaces were clean and ready for Monday morning. The forklift was at the dock door readied for moving in the next igloo or metal pallet after Monday's pickup. He thinks the Monday load will be from the Air Force; they have some metal pallets to be sandblasted and cleaned. The bins for repair tickets were full of blank forms; there were a few boxes left over with some netting material in them and a packing slip to check once more before mailing; the sewing machine and strapping material were wrapped to protect them from the dust. Everything looked spotless and clean.

The repair orders were coming in regularly now, but they still only used one shift, 5 days a week. With their staff of college students, Harman was repairing from two to four cargo nets and straps per day, performing sand-blasting, cleaning, and minor repairs on two Air Force metal pallets per day, and fixing about five igloos from commercial cargo carriers every few months. The record keeping was simple, using a mixture of Excel spread-sheets to keep track of the different nets, pallets, and igloos repaired. They kept track of brand name, the owner of the item (usually just the Air Force), and two to four air cargo carriers. The bin of common replacement parts stayed full enough to cover a week's worth of work: metal fasteners, thread, all sorts of metal pins, and different toxic paints. The hazmat drums were taken away about once every 3 months, filled mostly with old paints or cleaning materials. Although the business seemed to be moving smoothly, and cash was flowing in, there was a concern Barry had. He thought that he could double the work they had with some new contracts, especially for the air cargo igloos. That was what he wanted to sell to Don: how to increase repair sales to the major cargo carriers by using RFID. Of course, one major problem was to find the right performance metric to use to sell Don the idea, and then for Don to sell the airlines. Barry had told Don over the last few months that RFID could be a way to improve business.

The sell

Don came in a few minutes later, a box of doughnuts in hand, ready to settle in, checking over last week's work orders, and hearing what Barry had to say. "So, what's all this about RFID, and how's it going to change our business?" asked Don, as he sat down at the former dining room table, now a conference table.

Barry said, "We have a good system at work, making money and doing a good job for the Air Force and a little work on the commercial side. But we can double our business if we just start using RFID tags to track the igloos we're repairing." Barry went on to explain how he had heard about the airlines were investing in RFID tags to track luggage, and the pilot studies would be completed very soon. Additionally, from his friend in baggage handling, he had overheard three airline executives talking about getting ready to invest several million to expand into RFID research in other areas. The airlines, though, were not sure in which area to invest other than luggage. The summer tourist season was ending, so that would give them time to get ready for next summer and convince the airlines to invest some of that money into Don's repair business. Barry said, with a little checking, he found out that the airlines ship their igloos to China for repair. The few that come to Don's shop were those that need instant repair, those that cannot wait for shipment to China. "But there are over 500 igloos across the street at the airport at any one day, and we get over 600 widebody jets in here weekly," Barry said. Barry continued that with the large amount of cargo movement at the airport, an opportunity was just there for the picking, even if the air carriers thought they could get cheaper repairs in China, rather than here in Anchorage.

Don asked how using RFID tags could possibly make them money, because the nets, pallets, and the few igloos that came in only had an order number to identify them for repair. There were no bar codes, even. In fact, the repair shop had no facility to use bar codes, so why upgrade to some wireless tag technology, which was supposed to be the future beyond bar codes? The manual method of tracking and tracing the pallets and nets and igloos inside the warehouse seemed to be working just fine.

Barry tried to explain that if they were to use RFID tags on the igloos, the minute they came into the warehouse, they could code the repair order number, the airline that owned the container, the type and make of the igloo, and the type or category of damage to the igloo. As the igloo moved through the warehouse, they could also update the passive RFID tags or their own database with the type of materials used to repair the igloo. There would be an accurate time stamp for when the igloo entered the warehouse from the receiving dock and the time it left by the shipping dock door. He said all they would need is an antenna and reader at the two dock doors. Barry said, "While we move the igloo around the warehouse for repair, they do not need to do anything but update the inventory items used for repair and record the workload time for the employees." Don replied, "I still don't see the need for these RFID devices. I know all about them from a magazine I picked up the other day, and I know how great they seem to be at saving retail stores like Wal-Mart on inventory items, but we are a small shop. So, why do we need RFID?" Don was interested, however, in the possibility of increasing the size of the repair business with the addition of the igloos from the airlines. Moreover, yes, he understood that the airlines were investigating passive RFID tags for luggage, because so many hundreds of thousands of bags are lost each year. Stretching from lost luggage to the luggage carriers, the igloos, was something else. He could see no connecting thread.

When the igloos are damaged, they put them aside and ship them to China. Some few find their way to Don's repair shop. Barry said that if the airlines were investing in RFID technology, it would not be too far a stretch to track passengers, luggage, and the luggage and cargo containers. They just needed to be sold on the idea. In addition, Don's repair facility was a perfect place to demonstrate this capability. The repair shop was right across the road from the runways and hangers. The igloos would come in, be tagged, and then leave with a permanent license plate of information that the airlines could then track for themselves. "We would do the initial tagging of all their igloos, as well as offer to provide a history of types of igloos used and that needed repair," he said. The airlines would reap a benefit in data mining on igloo conditions. Also, if the airlines were to place RFID readers in all their airports, they could track and trace each igloo and cut down lost igloos, which seems to be an industrywide problem. Therefore, Barry said, "All we have to do is convince the airlines to let us tag their igloos." Don still looked skeptical. "I am still not convinced that we need RFID tags," Don said.

Case analysis

What do you think the problem is for Don with the use of RFID for igloos? What false assumptions, if any, did Barry make in trying to sell RFID to Don? How could Barry have sold this idea any better to Don? What advice would you give Don on his company's investing in RFID?

Case study 2

Muddy boots and smart wood

There was no question that the weekend was starting out as one Taylor would rather not be at work. However, she was the purchasing director of Vapid Lumber Industries (VLI), and the inventory had to be counted, one stick of lumber after another. The day was gray, the rain had started, and the workers would not show up today. This was the only time the lumber-yard was quiet, the saws were not running, the forklifts not racing around, the fans not blowing, and the boss not screaming at some worker for not creating the door joints fast enough. Taylor's job was to walk the yard every 30 days, rain, sleet, hot, or cold. Today, in the springtime, the yard had more than nearly twice the lumber as last month. Bill, the sales representative, had been working very hard the last 2 months to stack orders, because he wanted a large bonus for his summer vacation coming in a few months. Then Taylor would do purchasing and sales and inventory together. Taylor went into the trailer office to get her pad of paper and pen. The yard was already ankle-deep in mud from the forklift and the flatbeds running around for the last month, and it seemed to be raining for the last 30 days.

The week before, Bill had been complaining that it would make his job and her job easier if they had a bar code system to read the inventory. Last year, the lumber distributor had starting stapling bar codes on the ends of each long and short piece of wood. But that was doing them no good at VLI, where the boss, Bob, would not hear of it. Bob was always trying to find ways to cut corners on the job. It took Taylor 2 years to talk Bob into buying a computer and convincing him that the Internet and simple spreadsheets could help him see the cash flow and flow of products and problems around the plant. That was enough technology for him, although Bob did seem to appreciate the monthly spreadsheets of how much wood was being delivered late from several distributors, and how the inventory and waste wood was fluctuating. That caused a few men to lose their jobs at first, but the use of the computer had not eased the inventory counting process or the inventory sheet written by hand that had to be sent to the home office in Texas each month. Taylor and Bill tried many times over the last year to convince the people at the home office that having 230 inventory sheets faxed to them each month could be replaced by sending the same information by Internet.

The big idea

The lumber waiting for cutting and shaping into door frames, window frames, and other construction special orders was sitting outside, in the mud. There were 12 rows of lumber, each stacked about 10 feet high, and about four loads per row. As Taylor walked around, drinking her morning coffee, she decided to simply walk the yard and see what could be done with the use of bar code readers or RFID. Stuffing the pad and pencil into her jacket pocket, she examined each of the bar code labels on the ends of the wood just delivered. It was starting to rain again, and the ground fog was still around. The sky showed no promise of today's weather getting much better. As Taylor examined the bar codes, she noticed some of them were clean and readable, but as she rounded the corner to those stacks of lumber that had been in the yard for a week, she noticed some of the tags were torn. Must be either the manhandling or forklift driver running into things again. Or could it be the rain? She felt a few of the tags; they were definitely soggy and could easily be torn or scraped as someone bumped into the ends. Still, Taylor thought, if we could use a bar code reader for each flatbed load as it arrives, she would have an instant count of what was arriving, instead of having to count each one on the flatbed before she had it offloaded. Also, if she had a portable reader, she could probably be using one now as she walked the yard. It looked like most of the lumber had bar codes on the ends; probably about 10% seemed damaged or were just missing.

So why not tell Bob to buy a bar code reader? That way, he would only have to pay for her time to walk quickly around the yard, rather than having to do it three times, as the company demanded she do. Counting the same thing three times was boring, and it was cold and wet, and it took all day, and sometimes most of Sunday.

As she was coming around the end of the lumber stack, she noticed a pickup truck. It was Bob. He never came in on Saturday. As Bob came over, Taylor decided to tell him her idea for using bar codes instead of hand counting the inventory.

Bob said, "You have what sounds like a good idea, Taylor, but the boys in Texas want an eyeball count. They don't trust technology. They trust you to see and count what is really out here in the yard. And that's what they

want done at the other 229 lumberyards today, all over the United States."
Bob went on to explain that the home office had been burned in the past
with computer technology. Also, he said, this was a hands-on operation, a
very simple manufacturing job. You take rough wood, cut it down, and make
builder-grade door frames, window frames, door blocks, and pallets. The
manufacturing process was simple, and simple cost less money. That was
the theory, and that was what Texas wanted and that was what they would
get. Bob then left Taylor to go into the office to pick up some papers, and
he was gone in a few minutes.

Taylor just stood there, in the rain, looking at her watch. It was 7:15 a.m.
She would be here until at least 5 p.m. How to convince the management here
and at the home office that technology could be useful was the question she
pondered, as she finished her coffee and began counting boards—one at a time.

Case analysis

What should Taylor do to convince Bob that the use of bar codes could be
   helpful?
Should Taylor and Bill go to Bob with an even more outlandish idea, such
   as RFID?

Case study 3

Alaska Supply Chain Integrators' cost of goods

Alaska Supply Chain Integrators (ASCI) purchases goods for oil companies
working on the North Slope of Alaska. The North Slope is the oil production
field where crude oil is extracted and then transported to the shipping
terminals in Valdez, AK. From Valdez, the oil is shipped to other ports on
its way to becoming refined petroleum products such as gasoline.

ASCI purchases approximately 40,000 items per year for these oil companies. The cost of goods (COG) purchasing, handling, and transportation
process is subject to many variables.

Background

To enhance ASCI's supply chain capabilities, it has developed a
state-of-the-art supply chain management and electronic commerce tool—a
software system. This system facilitates control of the procurement process
through a series of checks and balances. The various software modules
describe the business functions in their names: SmartTracker, SmartCatalog,
SmartMarkets, SmartMeasures, SmartBOM, SmartSpecs, SmartActions,
SmartTagger, SmartBundler. Key to many of these e-commerce capabilities
are time-sensitive measurement metrics.

ASCI's time measurement metrical units include weekly, monthly, and
quarterly timeframes for three tiers of vendors. These internal metrics
are also linked to the Balanced Scorecard method to track vendor delivery

compliance agreements. Vendors are held accountable to on-time and accurate delivery for the items they provide. Through the use of these metrics, ASCI's 2005 performance measurement achieved 90.6% on-time and accurate delivery of goods to North Slope customers.

The problem

ASCI's e-commerce solution is intended to address the lack of visibility in tracking or capturing the movement of goods from vendor to end user or client. This lack of visibility has been identified as a recurring supply chain management problem that, if adequately addressed, would add value to the client.

To be fully inclusive and contribute the highest value, this visibility must provide not only tracking of goods purchased but the handling and transportation activities, as well. The logistics of moving the procured items provides a means to measure and enhance added value to the product while assisting ASCI in lowering the final price paid. As an example, one frequent event includes the misclassification of purchase price with the total price of goods purchased, which includes transportation costs. In this situation, the cost of transportation is buried in the cost of the goods' purchase price. The indicated cost of material is inflated by the transportation cost, which prevents or hinders the supply chain management team from identifying the true cost and potentially addressing root issues.

One key factor in providing the overall supply chain visibility is the capability to track the product from receipt to final delivery. Currently, this tracking is achieved through a combination of bar codes, visible inspection, and person-to-person contact augmented by manual computer entry.

The RFID solution

At a recent meeting with representatives of MX Consulting, a proposal was presented to show that RFID could save ASCI more time and money in tracking goods from the vendors to the end customer's warehouse. Meeting with ASCI manager, Scott Hawkins, and several senior purchasers, Bob Tibmen, president of MX, explained, "With passive RFID tags on each high value item, like the generator, you could track and trace its movement from the factory to the North Slope warehouse." Scott and his staff listened as Bob laid out how an RFID system not only could provide visibility along the entire supply chain, but could also be very useful in ASCI's cross-docking facility. Bob explained that when pallets of goods entered the facility, they could instantly record what the product was, where it had come from, the due date to get to the North Slope, and any other related information. The key to successfully using RFID is that when boxes of items come on a pallet, ASCI no longer would have to open each box and read each item by bar code, or record any handwritten information that might have been added to the bar code label, as often happens. Bob said, "The use of an RFID tagging system could cut down on your labor cost; you could probably eliminate one or two positions in loading and processing and checking of the products as they enter your facility."

After a round of questions by the ASCI staff about the physical properties of RFID, Scott finally said he had to go, and "would call Bob in a few days with an answer."

Case analysis

What advice would you give Bob if you were one of Bob's employees?

What advice would you give Scott if you were one of Scott's managers or employees?

What is the basis for determining the price on the COG that could be changed by RFID?

What savings on the COG could be realized with an RFID system?

Case study 4

Radio chips in credit cards

Gas station chip kill

Sara Wallace opened her mail and found her new company credit card. Sara was the operations director of MOAT Transport, LLC, and was excited to get her new business card. She was on her way from the office to the gas station to fill up the company truck and go check out the new cross-docking warehouse.

Although Sara was excited about the new card, she was skeptical about using it, because it was one of those new Scan-N-Go cards, containing some kind of radio frequency chip. Sara got out of the truck, turned to the gas pump, and held out the credit card in front of the Scan-N-Go scanner. The scanner immediately beeped and the small computer screen displayed the message, "Sara, welcome to Strickland's Gas Station. Would you like a car wash today with your purchase? Push Yes if you would."

Sara looked at the screen for a few seconds, looked at the card, and then placed the card on the ground and stomped on it with her work boots. Later, she told her boss, Will, "I stomped on it 10 times, then held it up to the scanner." The gas pump was silent; the computer screen was silent. Satisfied, Sara told Will that she then used the credit card in the card reader the way she had always done. She said, "I killed the card. Well, I killed that radio chip inside."

RF dollar block

Ashley had been reading the magazine and newspaper accounts of these RFID tags that would be used inside your credit and debit cards. She read that soon every card would have one of these little chips embedded inside. The reports of these smart cards were that you could still use the cards with a card reader as before, but now all you had to do was pass the card in front of a scanner or RFID reader and your purchase would be made automatically.

Ashley had also just received her new passport. She was not very pleased to now have one of those RFID chips inside her passport any more than she was about credit cards being able to be read by some simple radio frequency device. She had recently run across a story on the Internet about companies that were against RFID, which pleased Ashley. She wanted to do something besides just be against RFID.

Ashley also ran a small, yet profitable, Internet business, selling helpful hints to housewives stuck at home with the kids. Ashley was one of those moms stuck most of the time in the house raising her two very young girls. From her home in Virginia, Ashley had been selling small hand wipes for moms across the world, along with a recipe on how to make them at home. Ashley had many helpful cost-saving tips. And, after 6 years of this small business venture, she had a large following of mothers across the world, from the U.K. to Australia and all across the United States.

Ashley wondered if these other mothers felt the same way she did. So, without much thought, she wrote a note about how she felt and sent it to her mailing list.

By nightfall, after the kids were in bed, Ashley again looked at her computer. What she saw was amazing. There were over 500 e-mails waiting to be read with the same reference subject, "What about these RFID tags?"

Ashley stopped reading and responding to the e-mails shortly after midnight; she had to get some sleep. But the response was overwhelming.

By morning, Ashley had another of her brilliant, yet untried ideas. She would cut a piece of aluminum foil, add some stickers on it, laminate it, and stick it in her wallet. She had read that her passport had a similar foil lining to stop people from unauthorized reading of passports. So, why not make a fake dollar bill-sized foil and place it in your wallet? It would fold around your credit cards and no one could read it, except when you took it out of your wallet to use. Would it work? Ashley did not really know.

Ashley decided to post this idea to her e-mail friends, and she would sell it for $3.00. So, she spent a minute taking a digital picture of her creation, posted it in a new flyer, and sent it out to all on her e-mail list.

The rest of the story is history. After only 1 month, Ashley is receiving over 100 envelopes per day in the mail, each with $3.00, $6.00, or $9.00. A few have $15.00 in each envelope.

Case study analysis

What kind of RF safety and security safeguards should be built in to credit cards, passports, and other personal identification tags?

What are the major credit card companies doing to keep the use of RFID chip credit cards from being seen as a threat?

How could criminals exploit this fear of credit cards with RFID chips?

Case study 5

Cool chain disaster

"We just turned away 4000 pounds of fresh wild Alaskan salmon," complained Tony, purchasing agent for Platt Seafood, Boston. "That's 40 boxes of garbage, a week's supply that our customers are screaming about, and now I have to place another rush order." Tony was very angry. This was the second shipment since May that a large order of seafood had arrived spoiled. This time, it was too obvious. After 30 years as a fisherman, fishmonger, and now purchaser for his own retail chain, he knew when a hot load had arrived. In addition, this load had all the signs: He did not have to do an enzyme test; he could tell by touch and smell. This fish would not last a day. "Order refused," Tony shouted over the telephone to his supplier in Kodiak, AK.

Kodiak

Olivia was probably as angry as Tony had been when she finished talking and listening to Tony. This was not good for her Cold Water brand of fresh, wild-caught Alaskan salmon. Tony had just canceled his contract with her for any more seafood of any kind from Cold Water. Olivia had taken every precaution to get this latest shipment to market. She had to think. What had gone wrong?

She remembered this load because it was the best-handled load, using seasoned crews, not the part-time college kids from Anchorage. The fish had been iced right away on the boat, and this crew wasted no time at all at the dock packing the boxes for flying out that same day. The truck was waiting just for this load and the short ride to the airport.

Amy had ridden along with the load to make sure the load was put onto the GenAir cargo flight. The plane had taken off within 2 hours of getting the load to the airport. One hour later it had landed at Anchorage. There had been a Cool Commodity Freight (CCF) truck ready to take the load to the warehouse for repacking. So, within 2 hours of leaving Kodiak, the entire shipment was being put into the cool storage warehouse at CCF.

Olivia had called Tia at CCF to make sure her load was being handled properly, and Tia had assured her it was. It would be iced and packed into the insulated igloos for the next day, when it would leave Anchorage on its way to Boston. Olivia checked her Internet logs and saw that the shipment had departed Anchorage on time the next day and had arrived in Chicago, where it changed flights. That took about 4 hours, and then on to Boston. The flight logs to Boston showed that the plane's cargo arrived right on time. And she knew that Tony was there to pick it up right at the Boston Cool Chain warehouse down from the airport. So, what could have gone wrong? The timeline for transportation was right. The fish had the proper ice level. What could have caused the fish to be so bad when it got to Tony?

Chicago

The GenAir cargo plane arrived on one of the hottest days of August. The temperature was over 100°F, and on the runway it was even higher. Robie had overseen the cargo offloading from Alaska as usual. But today was to be a special problem. When he went into the cargo hold, it was a mess. Water everywhere. Boxes with water dripping all over the place. And the heat and humidity was really oppressive. It was like an August storm day in Louisiana. It would take an extra cleaning before this plane could be turned around.

He noticed the fish boxes right away. They were hot to the touch. This was not a good sign. GenAir had a great record of on-time delivery of seafood from Alaskan processors, and today's load was going to become a problem if he did not do something. He called his supervisor. Robie said, "What should I do with this load from Alaska? It feels hot, and I know it is." His supervisor, Henry, glared at Robie and said, "Load it on the next plane, and then get the cleaning crew and maintenance guys over there now. We have to turn this plane around in 6 hours for the flight back to Alaska." So Robie had the 40 boxes loaded and moved to the Boston plane. He knew this was bad; it sat in the hot sun for the next 4 hours. Then it was loaded and on its way to Boston, out of their hands, not in Chicago, not on GenAir. Robie had called the maintenance supervisor, Evan. Sure enough, there was a little thermostat problem, fixed in about 15 minutes.

Case study analysis

Think through the seafood supply chain of catching and moving the seafood cargo from boat to the customer in Boston. What were the trouble spots along this supply chain? Explain your answers.

What assumptions were made along this seafood supply chain by the various stakeholders? What were any vulnerabilities of these assumptions?

Could RFID time and temperature tags have helped this situation? Think carefully about your answer, from the boat to the customer. What are the implications, the direct and indirect costs of using RFID? What is the value to be added to shipments of seafood using RFID?

Epilogue

Radio frequency identification, or RFID, stories for the last few years have been numerous, and they are almost all about the same thing: Wal-Mart and the Department of Defense mandating their suppliers to use RFID tags for pallets and containers as a replacement for the standardized, accepted bar code. But there are continuing reports of potential warnings or pitfalls that RFID tags probably will not begin to replace bar codes for at least a decade.

There is a growing concern by many experts and journalists about this ubiquitous dark side of RFID, where supply chain corporate leaders will use them as "spy chips."[1] This may be a benefit to helping secure the country's borders from illegal packages entering the country. But as a spy chip, others see civil liberties at risk. This fear of misuse of RFID technology is compounded by the increasing numbers of experts or managers who still do not know enough about RFID or what the RFID issues are that face supply chain or logistics firms.[1] "No one knows how RFID tags will perform in extreme conditions."[2] No one knows what to do when the tags do not work.[2]

Economic or operations research studies have not yet put a definitive return on investment with RFID, which further fuels the distrust of this technology insertion. "All this security stuff costs as much as the container itself."[3] "There is no return on investment for manufacturers."[4] Many trade stories and discussions with business leaders and students assume that replacing bar codes with RFID tags is not going to happen.

Besides the uncertainty and the lack of defensible metrics, there is a process gap in the standards needed to use RFID technology efficiently and effectively. "Lack of standards is the biggest challenge; the second biggest being the cost of the tag."[2] "In international logistics, RFID systems are today virtually non-existent."[2] "The most common objection is that RFID technology is not mature enough."[5] Wal-Mart does not have a pilot. Well, there are pilot projects abounding around the country; many firms are contracting with experts to experiment with RFID in their logistics systems; many are trying to grow their own experts in the technology in hopes of lowering these fears and risks, which mainly is the bottom line: Do we still make a profit with RFID?

With all these dire warnings, RFID technology seems to have created its own tidal force in transportation, materials management, and really all aspects of logistics and supply chain economy. To better understand this force, let us step back in time a decade, or four, to better see what may be happening to our lives with all this talk and energy and money being dumped into RFID tagging on pallets or products within the logistics systems from the mom-and-pop stores to Wal-Mart. Are we in another computer revolution of some kind, or another event as traumatic as Y2K? Alvin Toffler described the Industrial Revolution as a "flash flood in history ... where many streams of change flowed together to form a great confluence."[6] The current wave of RFID appears to be part of yet another flash flood of technology that will alter the computer industry one more step toward some as-yet-defined global communications system of systems.

But having lived through the last 4 decades of how the computer, and a soup of acronyms from ADP to artificial intelligence (AI) to knowledge management (KM), has changed the modern office and workplace, how can we measure the future impact of RFID, which promises to deliver billions of pieces of data, hourly, to our workplace? Toffler went on to say that "any search for the cause of the Industrial Revolution is doomed" because there are so many possible causes. He did go on to say that "technology, by itself, is not the driving force of history."[6] This is important because there are many articles being written about RFID technology from the perspective of the technology itself, mainly its limitations. Toffler warned us that because we cannot find the single cause of technology of industrial impacts on our society and culture, "the most we can do is to focus on those that seem most revealing for our purposes and recognize the distortion implicit in that choice."[6] So far, these distortions seem to be people warning of excessive cost or how the technology is not quite ready.

We may begin to glimpse the future impact of RFID technology when we remember Toffler's concerns that "the greater the divorce of producer from consumer ... the more the market ... with all its hidden assumptions ... came to dominate social reality."[6] In the 1980s, he was concerned with the limited nature of the communications systems between the manufacturers and retailers in the long supply chains that separated the customer from what the customer really wanted in a product. Now, in the 2000s, that has changed, with the advent of just-in-time manufacturing, Internet, B2B, and other business model changes, focused solely on the customer's near-instant demands. All these warnings about the use of technology form a useful pattern of pitfalls that we should watch for as we rush into adopting RFID technology. Besides the warnings at the beginning of this chapter, one of the earliest signs of a huge pitfall is from the antitechnology sector of the workforce or customer base.

There has always been an antitechnology force at work in our society, since the early days of the Industrial Revolution. There are decades of studies on the causes and identities of those in the antitechnology movements. However, one of the early technology watchers examined the most recent example, in our memory, of technology changing our way of life. H. W. Lewis said that "displaced workers do not form the core of the antitechnology movement ... it seems to be an upper-middle-class phenomenon."[7] When you read the majority of the trade magazines in the last year about RFID, you get the sense that the workingman or -woman in the warehouse or the truckers will be against the use of the technology. However, that does not seem to be the case.

At the beginning of 1990, John Naisbitt said, "Telecommunications and computers will continue to drive change, just as manufacturing did during the industrial period."[8] Naisbitt went on to say, "We are laying the foundations for an international information highway system ... moving to a single worldwide information network ... becoming one global marketplace."[8] This appears to be what will finally happen with the advent of widespread adoption of RFID, but it comes at the price of a crisis of acceptance of this newest paradigm, or model, of technology replacing the older paradigm: bar codes.

Thomas Kuhn said, "Crises are a necessary precondition for the emergence of novel theories."[9] Although he was talking about scientific theories being replaced, his concepts are equally applicable to our everyday technology acceptance or rejection. He also said, "To reject one paradigm without simultaneously substituting another is to reject science itself." That act reflects not on the paradigm but on the person. Inevitably, their colleagues will see them as "the carpenter who blames his tools."[9] There are concerns from those who have invested heavily into bar codes; they will not go easily into RFID. The old tool of the bar code works; why mess with it? So, it is to be expected that over the next few years, we will see a flurry of articles evoking crises of confidence over the new tools for the data carpenters of logistics systems.

Kuhn said, "If an anomaly is to evoke crisis, it must usually be more than just an anomaly. There are always difficulties somewhere in the paradigm-nature fit; most of them are set right sooner or later, often by processes that could not have been foreseen."[9] So, the question to you and to the CEO or general manager contemplating the switch to RFID is, Can you identify these difficulties in time to avoid too many crises within the workforce, or in the customer base?

Kuhn said, "Often a new paradigm emerges, at least in embryo, before a crisis has developed far or been explicitly recognized."[9] Many articles cite the immature nature of RFID; they see the technology as some embryo and not ready for capital investment.

On the benefit of bar codes or RFID, Kuhn helps us further understand what is happening. He said, "When paradigms enter, as they must, into a debate about paradigm choice, their role is necessarily circular. Each group uses its own paradigm to argue in that paradigm's defense."[9] Thus, when reading about bar codes versus RFID, the very theme pitting one against the other is not a sign of the failure of RFID; it is a stage-setting premise, a dance, if you will, that is being played out in the natural course of technology paradigm changes. One technology is changing, and the owners are reluctant to let go; not for bad science, but just because that is the nature of paradigm shifts.

But there are those who insist on facts to know how good RFID is. How do we measure the difference between bar codes and RFID? Kuhn again helps us understand, when he said, "The competition between paradigms is not the sort of battle than can be resolved by proofs."[9] He also said, "In the first place, the proponents of competing paradigms will often disagree about the list of problems that any candidate for paradigm just resolves. Their standards or definitions of science are not the same."[9]

When we go from bar code standards to RFID standards, as has been indicated in the press on this matter, there is "more involved than the incommensurability of standards. Since new paradigms are born from old ones, they ordinarily incorporate much of the vocabulary and apparatus, both conceptual and manipulative, that the traditional paradigm had previously employed. But they seldom employ these borrowed elements in quite the traditional way. Within the new paradigm, old terms, concepts, and experiments fall into new relationships with the other. The inevitable result is what we must call, though the term is not quite right, a misunderstanding between the two competing schools."[9]

So, how and when will this transition from bar codes to RFID take place? Kuhn helps us when he said, "Before they can hope to communicate fully, one group or the other must experience the conversion that we have been calling a paradigm shift. Just because it is a transition between incommensurables, the transition between competing paradigms cannot be made a step at a time, forced by logic and neutral experience. Like the gestalt switch, it must occur all at once (though not necessarily in an instant) or not at all."[9] What this means is that what Wal-Mart is doing is following the models of paradigm shifts set out by Thomas Kuhn from 30 years ago.

Again, when does transition from bar code to RFID occur? Kuhn said, "The transfer of allegiance from paradigm to paradigm is a conversion experience that cannot be forced."[9] Yet, Wal-Mart and DoD are forcing suppliers to comply. Will their force wave be successful? Perhaps Wal-Mart and DoD have the answer. So, when we are asked by someone why we should replace bar codes with RFID, Kuhn once again helps us understand. He said, "Probably the single most prevalent claim advanced by the proponents of a new paradigm is that they can solve the problems that have led the old one to a crisis."[9]

So, what is this word *paradigm* that we have used? According the Kuhn, "The term 'paradigm' is used in two different senses. On the one hand, it stands for the entire constellation of beliefs, values, and techniques, and so on shared by the members in that constellation. On the other hand, it denotes one sort of element in that constellation, the concrete puzzle-solutions which, employed as models or examples, can replace explicit rules as a basis for the solution of the remaining puzzles or normal science."[9]

For those who are still skeptical, as these pitfalls indicate you could be, John Naisbitt stated in his best-selling book that "you may choose to challenge the trends, but first you must know where they are headed."[8] That RFID is here to stay is a fact. That bar codes are here for quite some time yet is also a fact. However, the smart investor will look beyond the emotion of the posturing by the old paradigm advocates and the antitechnology purists, and realize that this is just another step in our long climb into the computer age, which is still defining itself.

References

1. Hickey, Kathleen, Loose chips sink ships, *trafficWORLD*, 12, 14, 2004, accessed by Internet December 6, 2005.
2. Hickey, Kathleen, Cold feet on RFID, *The Journal of Commerce*, 32–33, 2003, accessed by Internet January 26, 2006,.
3. Hickey, Kathleen, RFID or bust, *trafficWORLD*, 20–23, 2004, accessed by Internet January 31, 2005.
4. Hickey, Kathleen, Wal-Mart's tall order, *trafficWORLD*, 9, 2003, accessed by Internet April 19, 2004.
5. Dunlap, Joe, RFID misconceptions revealed, *Parcel Shipping & Distribution*, 26–27, 2003.
6. Toffler, Alvin, *The Third Wave*, Bantam Books, New York, 1980.
7. Lewis, H.W., *Technological Risk*, Norton, New York, 1990.
8. Naisbitt, John and Aburdene, Patricia, *Megatrends 2000: 10 New Directions for the 1990s*, Avon, New York, 1990.
9. Kuhn, Thomas S., *The Structure of Scientific Revolutions*, 2nd ed., The University of Chicago Press, 1970.

Index

A

Active RFID tags, 13–15, 47; *see also* RFID
Adam Smith, 47
ADP, *see* automatic data processing (ADP)
Agriculture, applications in, 5, 48, 50,
    51, 60
AI, *see* artificial intelligence (AI)
Alaska, RFID usage, 51–53
Algorithms, *see* modeling
Alien™ RFID tag, 5, 59
Animal implants, 4, 18, 44, 48
Antitechnology movement, 83
Applications of RFID, 4, 10, 17, 18–19, 43–46;
    see also usage of RFID technology
        access control, 44
        agriculture, 5, 48, 50, 51, 60
        cargo shipping, 15, 18, 48
        hazardous waste transportation, 48
        in libraries, 45, 51
        luggage tags, 10, 70
        mail, 45
        security, 17, 18
        tamper-resistance, 68
Artificial intelligence (AI), 1, 16, 20, 87
Assumption-based planning (ABP), 73–75,
    80–81; *see also* war games
Auto-ID systems, 4, 19–20
Automatic data processing (ADP), 1–2

B

Balanced Scorecard, 23
Bar codes; *see also* smart labels
    accuracy, 89–91
    adoption of, 16
    costs, 14
    data, 7–8, 9
    durability, 9
    label-reading process, 5–7, 9
    replacing with RFID, 9, 84
    versus RFID tags, 3, 4–9
    used with RFID tags, 10
    variety, 28
Biometrics, 19, 51
Border security, 17, 19, 44, 68, 80
Born-on-dates, 29
Brown, Steve, 12
Business plans, 75, 76
Business re-engineering, 3, 16

C

Cardullo, Mario, 4
China, RFID adoption, 46, 47–49
Closed-loop systems, 24–27, 39
Communications
    and closed systems, 25
    and customer service, 35
    global, 77
    and the Internet, 83
    with RFID, 22
    technology, 3, 33, 57
    and transportation, 68, 70
Computers
    chips in RFID tag, 3
    expert systems, 16
    software for RFID, 10, 11
    technological development, 1–3, 4,
        31–32
Computer simulation, 40; *see also* modeling
Container shipping, 15, 18, 44, 48, 68;
    see also transportation
Conveyor systems, 17
Corrective action loop, 96
Cost of goods (COG), 47, 49, 61, 64

OTHER TITLES OF INTEREST BY TAYLOR & FRANCIS

RFID in Logistics: A Practical Introduction
Erick C. Jones, University of Nebraska, Lincoln, Nebraska
Christopher a. Chung, University of Houston, Texas
ISBN: 0849385261

RFID Devices Handbook: Technology, Design, and Applications
Akshay Tyagi, Quest Communications, Highlands Ranch, Colorado
ISBN: 0849332672

RFID in the Supply Chain: A Guide to Selection and Implementation
Judith M. Myerson, IT Consultant, Philadelphia, Pennsylvania
ISBN: 0849330181

For Product Safety Concerns and Information please contact our EU
representative GPSR@taylorandfrancis.com Taylor & Francis Verlag GmbH,
Kaufingerstraße 24, 80331 München, Germany

Printed and bound by CPI Group (UK) Ltd, Croydon, CR0 4YY
08/05/2025
01864361-0001